Mechanical Metamaterials for Engineering and Architectural Applications

Mechanical Metamaterials for Engineering and Architectural Applications examines materials with extraordinary mechanical properties such as negative stiffness, programmable elasticity, and auxetic behavior that surpass conventional limits. This accessible yet comprehensive book covers design principles, fabrication techniques, and practical applications, including vibration isolation in aerospace systems and adaptive building façades. Bridging theory and practice, it highlights the transformative impact of mechanical metamaterials across multiple industries.

Key features:

- Overviews of the design principles, fabrication techniques, and functional advantages of mechanical metamaterials
- Discussion of extraordinary mechanical properties such as negative stiffness, programmable elasticity, and auxetic behavior
- A forward-thinking vision for industry and research integration
- A focus on architectural and civil engineering applications
- In-depth exploration of functional behaviors addressing real-world challenges

This book is intended for graduate students and researchers in mechanical and materials engineering, civil and structural engineering, and architectural engineering.

Mechanical Metamaterials for Engineering and Architectural Applications

Kaushik Kumar, Nabanita Saha,
P.S. Balaji, and Chikesh Ranjan

CRC Press
Taylor & Francis Group
Boca Raton London New York

CRC Press is an imprint of the
Taylor & Francis Group, an **informa** business

Designed cover image: Kaushik Kumar, Nabanita Saha, P.S. Balaji, Chikesh Ranjan

First edition published 2027
by CRC Press
2385 NW Executive Center Drive, Suite 320, Boca Raton FL 33431

and by CRC Press
4 Park Square, Milton Park, Abingdon, Oxon, OX14 4RN

CRC Press is an imprint of Taylor & Francis Group, LLC

ISBN: 978-1-041-21837-1 (hbk)
ISBN: 978-1-041-32628-1 (pbk)
ISBN: 978-1-003-78367-1 (ebk)

DOI: 10.1201/9781003783671

Typeset in Times
by SPi Technologies India Pvt Ltd (Straive)

Contents

About the Authors

Kaushik Kumar holds a Ph.D. in engineering from Jadavpur University, India, an MBA in marketing management from Indira Gandhi National Open University, India, and a Bachelor of Technology from the Regional Engineering College (now the National Institute of Technology), Warangal, India. For 11 years, he worked in a manufacturing unit of global repute. He is currently working as a Professor in the Department of Mechanical Engineering, Birla Institute of Technology, Mesra, Ranchi, India. He has 24 years of teaching and research experience. His research interests include composites, optimization, non-conventional machining, CAD/CAM, rapid prototyping, and quality management systems toward product development for societal and industrial usage, and he has received 50+ patents for them, most of which are donated to startups and the industrial fraternity through the Jharkhand Small Industries Association (JSIA). He has published 60+ books (they are referred to as text books and reference books by 40+ universities and institutes in their academic curriculum), 100+ book chapters, and 200+ research papers in peer-reviewed reputed national and international journals. Kaushik has also served as editor-in-chief, series editor, guest editor, editor, editorial board member, and reviewers for international and national journals. He has been felicitated with many awards and honors, including the Distinguished Alumnus Award for Professional Excellence 2023 under Academic and Research from the National Institute of Technology, Warangal, India. He has also received sponsored research and consultancy projects of more than 1 crore from the Government of India and abroad. Kaushik has delivered expert lectures as a keynote speaker at international and national conferences, and as a resource person at various workshops, FDPs, and short-term courses. He has guided many students of doctoral, masters, and undergraduate programs of his home and other institutions in India and abroad. He has also served as a reviewer and examiner of doctoral and master's dissertations for institutes in India and abroad.

Nabanita Saha is an architect and urban planner with a distinguished academic background. She graduated from the Sir J.J. College of Architecture and pursued her post-graduate and doctoral studies at the Indian Institute of Technology (IIT) Kharagpur. Currently, she serves as an assistant professor in the Department of Planning and Architecture at the National Institute of Technology (NIT), Rourkela. Her research interests span globalizing cities, social networks, blue-green infrastructure, and spatial planning. In addition to her research, Dr. Saha teaches a wide range of subjects, including architectural design, landscape design, building materials, building services, building byelaws, urban design, and planning theory and processes.

Beyond her academic commitments, she actively engages in consultancy and sponsored research projects, contributing to architectural design, master planning, and community development initiatives. She has also explored the feasibility of utilizing unconventional stones in architectural applications.

P.S. Balaji is an assistant professor in the Department of Mechanical Engineering at NIT Rourkela, with a Ph.D. in vibration control from Curtin University, Australia. With over six years of teaching and research experience, he has expertise in vibration control, nonlinear dynamics, and drone technology. Dr. Balaji has been instrumental in securing major projects, including one on Drone Aeromechanics worth ₹1.55 crores, which led to the establishment of a fully functional lab equipped with high-class facilities for advanced drone research. In addition to his research projects, which are funded by prestigious organizations like SERB DST India and ISRO, Dr. Balaji has successfully supervised two Ph.D. students. He has also made significant contributions to academia, with numerous publications in reputed journals, authorship of conference proceedings, and book chapters. He holds two patents in the field of vibration control, showcasing his innovative approach to solving complex engineering challenges. Dr. Balaji is also a sought-after speaker, having delivered numerous talks at Faculty Development Programs (FDPs) and conferences, sharing his insights as a guest speaker.

Chikesh Ranjan is a highly accomplished mechanical engineer currently serving as a project engineer for Project SwaYaan—Capacity Building in Drone/UAS at the Department of Mechanical Engineering, NIT Rourkela. He holds a PhD in mechanical engineering from the Department of Mechanical Engineering, Birla Institute of Technology, Mesra, Ranchi, Jharkhand, India. Chikesh also possesses a Master of Engineering degree in mechanical engineering, specializing in the design of mechanical equipment, from the same institution.

His academic journey began with a Bachelor of Engineering in mechanical engineering from Marathwada Institute of Technology, Aurangabad, Maharashtra, India, where he graduated with first class with distinction. Chikesh's innovative contributions have been recognized through the publication of 15 journal papers, 10 authored books, and 13 book chapters. He has actively organized 10 workshops, coordinated 13 technical events, and been involved in 18 technical events overall. Additionally, he has participated in 31 training or FDP programs and attended 34 national and international workshops, seminars, and webinars.

Chikesh has been granted ten design patents and seven foreign patents, demonstrating his dedication to advancing mechanical design and technology. With nine years of teaching and research experience of global repute, his teaching and research interests include composites, non-conventional machining, CAD/CAM, and robotics. He is a proud member of the Institution of Engineers, India, and has received seven prestigious awards and honors for his remarkable contributions to the field of mechanical engineering.

Preface

The authors are pleased to present the book *Mechanical Metamaterials for Engineering and Architectural Applications*. The book title has been conceived as a comprehensive and structured resource introducing the emerging field of mechanical metamaterials (a common term for *Meta-Structured Materials*), a class of architected materials whose properties are primarily governed by their internal geometry rather than solely by their chemical composition. Unlike conventional materials, mechanical metamaterials exhibit unique mechanical responses—such as a negative Poisson's ratio, high stiffness-to-weight ratios, programmable deformation, and tunable wave propagation—arising from carefully designed micro- and meso-scale architectures. The aim of this book is to present the fundamental principles, design strategies, and engineering significance of these materials while highlighting how advances in computational modeling, topology optimization, and additive manufacturing have enabled the realization of complex material architectures.

This book systematically explores the design, mechanics, fabrication methods, and performance characteristics of mechanical metamaterials, including lattice structures, auxetic systems, hierarchical architectures, and other advanced architected materials. It discusses the underlying deformation mechanisms, structural optimization approaches, and experimental and numerical techniques used to evaluate their mechanical behavior. In addition, this book highlights the relationship between structural geometry and functional performance, providing insights into how metamaterial architectures can be engineered to achieve targeted mechanical properties such as enhanced energy absorption, vibration control, and adaptive stiffness.

Designed for graduate students, researchers, and practicing engineers in mechanical engineering, materials science, aerospace engineering, and related disciplines, this book bridges the gap between fundamental theory and emerging technological applications. By integrating foundational concepts with recent research developments and practical case studies, it aims to serve both as a teaching resource and a reference guide for those interested in the design and application of next-generation architected materials in fields such as aerospace structures, automotive systems, biomedical devices, and advanced structural engineering.

The entire book is divided into six **chapters**. The organization of this book reflects both a pedagogical and an application-oriented approach. To self-assess the extent of knowledge gained, each chapter includes subjective review questions.

The opening chapter, that is, **Chapter 1**, introduces mechanical metamaterials. Mechanical metamaterials are engineered materials whose mechanical properties arise from their designed internal architectures rather than their composition. By using periodic unit cells and lattice geometries, they exhibit unusual behaviors such as a negative Poisson's ratio and tunable stiffness. Advances in computational design and additive manufacturing have enabled their applications in aerospace, robotics, biomedical, and energy systems.

The next chapter, **Chapter 2**, elaborates on design and fabrication techniques. Mechanical metamaterials are designed using engineered architectures such as periodic unit cells and topology-optimized lattices to achieve tailored mechanical properties. Advanced computational design methods guide their development, while modern fabrication techniques, including additive manufacturing and microfabrication, enable the production of complex structures, resulting in lightweight, high-performance, and multifunctional materials for diverse engineering applications.

After fabrication, the mechanical properties and functional behavior have to be elaborated, so **Chapter 3** discusses these through various case studies from both domestic and industrial sectors. The mechanical properties and functional behavior of these materials are primarily governed by their engineered internal architectures rather than their base material composition. By designing specific unit cells, lattice structures, and periodic geometries, mechanical metamaterials can exhibit unique responses, including negative Poisson's ratio, enhanced energy absorption, tunable stiffness, and controlled deformation. These architected structures enable engineers to tailor material performance to meet specific functional requirements. Consequently, mechanical metamaterials offer improved capabilities, including vibration damping, impact resistance, efficient load distribution, and adaptive structural responses. When combined with advanced computational design methods and modern fabrication technologies, these materials offer significant opportunities for developing high-performance systems with customized mechanical properties and multifunctional engineering capabilities.

After fabrication and detailed evaluation of their mechanical characteristics and structural behavior, the next stage is to explore practical applications. **Chapter 4** focuses on the diverse engineering applications of mechanical metamaterials. These materials have attracted considerable interest because their unique performance stems from carefully designed internal architectures rather than solely the properties of the base material. Such structural design enables the development of lightweight systems with superior strength-to-weight ratios, efficient vibration mitigation, enhanced energy absorption, and multifunctional capabilities. As a result, mechanical metamaterials are increasingly used in fields such as aerospace, robotics, biomedical engineering, transportation, and advanced infrastructure, where high mechanical efficiency and adaptability are required. Their architected geometries allow precise control over properties such as stiffness, deformation response, acoustic characteristics, and thermal behavior. Combined with advances in computational design, additive manufacturing, and smart material technologies, mechanical metamaterials are emerging as an important platform for creating next-generation, high-performance, and energy-efficient engineering systems.

Recognizing the growing relevance of metamaterials in the built environment—and consistent with the focus of this book—**Chapter 5** is devoted entirely to architectural applications. In architecture, mechanical metamaterials offer new possibilities because their performance can be engineered through carefully designed internal geometries rather than relying solely on conventional material properties. Such architected materials enable the creation of building components that combine reduced weight with improved strength, while also offering benefits such as better thermal insulation, acoustic regulation, and vibration mitigation. When incorporated into elements like façades, walls, roofs, flooring systems, and structural frames, these materials can enhance overall building efficiency, durability, and energy performance. In addition, metamaterial-based designs can support advanced capabilities such as seismic resilience, responsive building envelopes, and adaptive structural systems that react to environmental conditions. Ongoing progress in computational design tools and additive manufacturing techniques is further enabling the practical realization of these concepts, opening new pathways for sustainable, high-performance, and multifunctional architectural structures.

For a comprehensive understanding of any subject, it is important to address both its limitations and its potential future directions. In line with this objective, the final chapter of this book, **Chapter 6**, discusses the emerging trends and challenges associated with mechanical metamaterials. These materials form a rapidly developing class of engineered systems in which unusual mechanical responses arise primarily from carefully designed internal structures rather than only from the base material itself. Recent progress in computational modeling, additive manufacturing, and smart material technologies has accelerated the development of metamaterials with programmable, adaptive, and multifunctional characteristics. Such materials show promising prospects in areas including aerospace engineering, biomedical technologies, smart infrastructure, robotics, and energy-related applications due to their lightweight nature and tunable mechanical behavior. Despite these advantages, several obstacles remain, such as large-scale manufacturing, long-term durability, complex design requirements, and the integration of multiple functional properties. Future research is expected to emphasize artificial intelligence–assisted design optimization, environmentally sustainable manufacturing processes, bio-inspired structural concepts, and the development of self-sensing or self-healing metamaterial systems. Overcoming these challenges will be crucial to enabling broader adoption and advancing next-generation engineering solutions that are efficient, adaptive, and sustainable.

The preparation of this book has been guided by the understanding that mechanical metamaterials are not merely advanced materials but a transformative approach to engineering design, where mechanical performance is achieved through carefully engineered internal architectures. By controlling geometry at multiple scales, these materials enable tailored properties, including tunable stiffness, enhanced energy absorption, and adaptive structural responses. By integrating fundamental mechanics, modern computational design methods, and recent advances in fabrication technologies, this book aims to contribute to the education, research, and technological progress in the field of architected materials. It is intended to support students, researchers, and engineers in understanding and applying mechanical metamaterials to develop innovative, lightweight, and high-performance engineering systems across diverse technological domains.

In conclusion, this book is a culmination of years of exploration, research, and personal growth. It is our hope that the insights and knowledge shared within these pages will inspire, challenge, and guide you in your own journey. Whether you are a seasoned professional or a curious learner, we believe that the ideas presented here will resonate and offer value. As you turn the pages, we invite you to engage with the material, question assumptions, and explore new perspectives. Thank you for allowing us to share this work with you—we hope it serves as a meaningful contribution to your endeavors.

Kaushik Kumar
Nabanita Saha
P.S. Balaji
Chikesh Ranjan

Acknowledgments

The completion of this book, *Mechanical Metamaterials for Engineering and Architectural Applications*, has been a deeply rewarding project, made possible through the generous support, guidance, and encouragement of numerous individuals and institutions. We would like to express our profound gratitude to all those who, in various ways, contributed to the successful realization of this work.

First and foremost, we would like to thank God for providing the opportunity to believe in passion, work with dedication, and pursue meaningful goals. In the process of developing this book, it was realized how valuable the opportunity to contribute to scientific writing is for any researcher. This work would not have been possible without the strength, perseverance, and faith granted by the Almighty.

Our deepest gratitude goes to our families for their constant encouragement and understanding. Their patience during the long hours of writing and their unwavering belief in us provided the emotional support necessary to complete this project. Their words of encouragement and ability to remind us of the broader purpose behind this effort were invaluable, especially during times when the demands of research and writing reduced the time we could spend with them. They remained a source of inspiration and motivation throughout the process. Their thoughtful discussions and support helped shape many aspects of this book, and their belief in our vision made a significant difference. We hope that one day they will read this work and understand why so much time was devoted to research, writing, and reflection. We also extend heartfelt thanks to our parents and grandparents for encouraging us to pursue our academic aspirations.

We extend our sincere appreciation to the academic and research mentors whose insights and constructive discussions shaped our understanding of materials science, structural mechanics, and the emerging field of mechanical metamaterials. Their guidance, scholarly rigor, and encouragement have continually inspired us to maintain the highest standards of technical clarity and academic integrity throughout this manuscript. Our heartfelt thanks are also due to colleagues and peers in the fields of materials science, mechanical engineering, and architectural engineering, whose exchange of ideas enriched several aspects of this book.

We are particularly grateful to researchers and professionals working in advanced materials design, computational modeling, and additive manufacturing, whose contributions to the field of mechanical metamaterials have inspired many of the discussions presented in this book. Their work in architected materials, lattice structures, and multifunctional material systems has ensured that the content remains aligned with current research directions and technological developments. We also acknowledge the technical reviewers and subject experts who offered valuable feedback on early drafts of the chapters, leading to improvements in scope, clarity, and depth.

Special thanks are due to research scholars and postgraduate students whose curiosity, discussions, and enthusiasm for learning continually motivated us to refine our explanations and strengthen the book's educational value. Their questions and engagement represent the spirit of inquiry that this work seeks to encourage. We are equally grateful to professional societies, research organizations, and academic institutions that have advanced the discipline of metamaterials research through publications, conferences, and collaborative initiatives.

We wish to record our sincere appreciation to the publishers and the editorial team for their professionalism, patience, and careful attention to detail throughout the publication process. Their support has been instrumental in transforming this manuscript into a refined and accessible reference for readers in both academia and industry.

Finally, we dedicate this work to the researchers, engineers, architects, educators, and students striving to advance the field of mechanical metamaterials and architected materials. It is our hope that this book will support learning, inspire innovation, and contribute to the development of lightweight, high-performance, and sustainable engineering systems for future technological and architectural applications.

Kaushik Kumar
Nabanita Saha
P.S. Balaji
Chikesh Ranjan

1 Introduction to Mechanical Metamaterials

1.1 INTRODUCTION

Mechanical metamaterials represent a rapidly evolving class of engineered materials whose mechanical properties are primarily derived from their architected micro- and mesoscale structures, rather than from their base material composition alone. Unlike conventional materials, in which stiffness, strength, density, and deformation behavior are intrinsically linked and largely fixed by chemistry, mechanical metamaterials enable unprecedented control over mechanical response through deliberate geometric design. By tailoring internal architectures, often in periodic, hierarchical, or lattice-based forms, these materials exhibit unconventional and often counterintuitive properties, such as negative Poisson's ratio (auxeticity), ultra-high stiffness-to-weight ratios, programmable deformation modes, negative compressibility, and tunable wave propagation characteristics (Chen et al., 2020).

The emergence of mechanical metamaterials has been driven by advances in computational mechanics, topology optimization, and additive manufacturing technologies, which together allow the precise realization of complex geometries that were previously impractical or impossible to fabricate. This convergence has expanded the design space of materials from a composition-centric to structure-centric paradigm, enabling multifunctionality and adaptability within a single material system. As a result, mechanical metamaterials are increasingly viewed not merely as materials but as mechanical systems embedded within matter (Jiao et al., 2023).

From an application perspective, mechanical metamaterials have demonstrated significant potential across diverse engineering domains, including aerospace and automotive lightweight structures, vibration and noise control, impact and energy absorption systems, biomedical implants and prosthetics, soft robotics, and adaptive civil infrastructure. Their ability to tailor mechanical behavior spatially and temporally also aligns closely with emerging concepts in smart materials and intelligent structures, where responsiveness, reconfigurability, and performance optimization are critical.

This chapter introduces the fundamental concepts underlying mechanical metamaterials, outlining their historical development, classification, governing mechanics, and key design principles. It establishes the theoretical and practical foundation for understanding how geometry, scale, and material choice interact to produce novel mechanical behavior, thereby setting the stage for subsequent chapters that explore specific classes, modeling approaches, fabrication techniques, and advanced applications of mechanical metamaterials.

DOI: 10.1201/9781003783671-1

1.2 DEFINITION AND EVOLUTION OF METAMATERIALS

Metamaterials are artificially engineered materials whose effective properties are governed primarily by their designed internal geometry rather than by their chemical composition. By arranging unit cells in periodic or architected patterns, metamaterials can exhibit unconventional and tunable properties such as a negative Poisson's ratio, negative stiffness, and wave-manipulation capabilities that are not observed in natural materials.

Figure 1.1 illustrates the fundamental concept of metamaterials by demonstrating how engineered internal architecture governs overall material behavior. The figure begins with a representative unit cell, which serves as the basic building block of a metamaterial. This unit cell is deliberately designed with specific geometric features, rather than relying on material composition, to control mechanical response. When the unit cell is periodically repeated, it forms an architected structure or lattice, where geometry, topology, and spatial arrangement collectively determine the macroscopic properties. As shown in the figure, this architectural control enables the realization of engineered material properties such as negative Poisson's ratio, enhanced energy absorption, and tunable stiffness. The figure emphasizes that, in metamaterials, properties are driven by internal structure rather than chemistry, establishing metamaterials as materials-by-design that can exhibit behaviors unattainable in conventional homogeneous materials.

The evolution of metamaterials reflects a progressive shift from composition-driven materials to geometry-driven material systems, enabled by advances in physics, mechanics, and manufacturing technologies.

Initially, metamaterials emerged in the late 20th century in electromagnetics, where artificially structured media were designed to manipulate electromagnetic waves in unconventional ways. These developments demonstrated that material behavior could be engineered through sub-wavelength structural patterns rather than intrinsic material chemistry.

Subsequently, the underlying design philosophy was extended to solid mechanics, giving rise to mechanical metamaterials. Researchers recognized that carefully designed micro- and mesoscale architectures could produce extraordinary

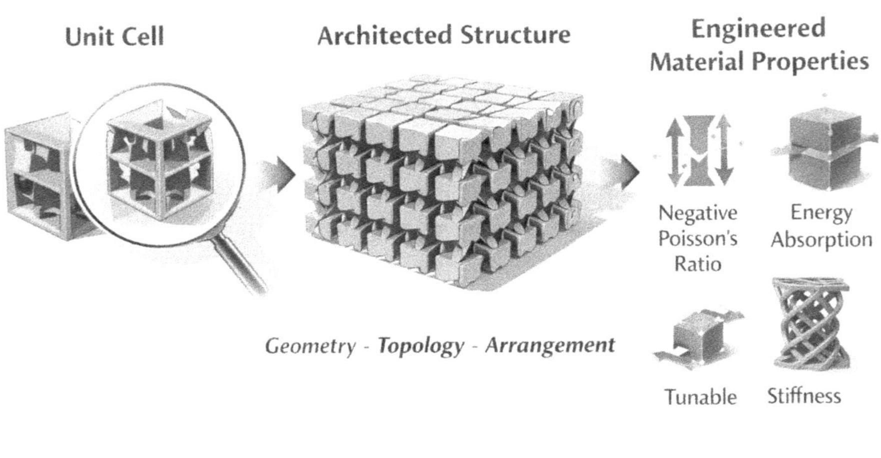

FIGURE 1.1 Conceptual representation of metamaterials.

mechanical responses such as negative Poisson's ratio (auxetics), negative stiffness, extreme energy absorption, and vibration bandgaps. This marked a departure from classical material limits defined by traditional elasticity theory.

In recent years, the evolution of metamaterials has been strongly driven by computational modeling, topology optimization, and additive manufacturing. These tools enable the precise fabrication of complex, hierarchical architectures and make metamaterials programmable, reconfigurable, and multifunctional. Today, metamaterials are actively explored for applications ranging from aerospace and robotics to civil infrastructure and adaptive architecture (Iyer et al., 2020).

Figure 1.2 illustrates the evolutionary progression of metamaterials from their conceptual origins to their present-day realization as advanced engineered systems. The evolution begins with electromagnetic metamaterials, developed primarily to control and manipulate wave propagation through artificially structured media rather than through intrinsic material composition (Reda et al., 2018). This foundational concept established the principle that geometry and periodicity at sub-wavelength scales can govern the effective behavior of materials. The next stage marks the emergence of mechanical metamaterials, where the same design philosophy was extended to solid mechanics. Architected lattice and cellular structures were introduced to achieve unconventional mechanical responses, such as negative Poisson's ratio and negative stiffness, thereby exceeding the limits of conventional materials. The figure further highlights the role of advanced computational design tools, including numerical modeling and topology optimization, which enabled systematic design, analysis, and optimization of complex metamaterial architectures. Finally, the evolution culminates in the era of additive manufacturing, in which high-resolution fabrication techniques enable the precise realization of intricate geometries, transforming metamaterials from theoretical concepts into practical, scalable solutions for engineering and architectural applications.

Table 1.1 summarizes the evolution of metamaterials from conventional composition-based materials to advanced architected systems governed by geometry and design. The progression highlights how initial wave-manipulation concepts expanded into mechanical systems and were further accelerated by computational design and additive manufacturing, positioning metamaterials as a key enabling technology for future adaptive and high-performance applications.

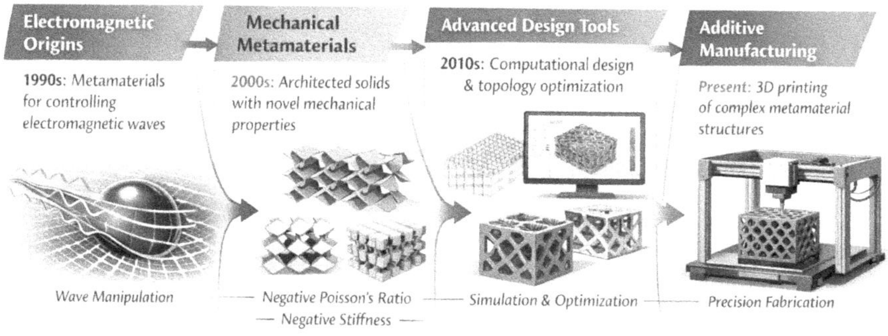

FIGURE 1.2 Evolution of metamaterials.

TABLE 1.1
Evolution of Metamaterials

Stage/Period	Focus Area	Key Characteristics	Description
Pre-Era (Before 1990s)	Conventional materials	Composition-driven behavior	Material properties were governed primarily by chemical composition and microstructure, with limited scope for tailoring beyond alloying and processing.
Electromagnetic Metamaterials (1990s)	Wave manipulation	Sub-wavelength periodic structures	Artificially engineered structures were developed to control electromagnetic waves, introducing concepts such as negative refractive index and establishing geometry as a governing factor.
Early Mechanical Metamaterials (2000s)	Architected solids	Geometry-driven mechanical response	The metamaterial concept was extended to solid mechanics, enabling unusual properties such as negative Poisson's ratio, negative stiffness, and enhanced energy absorption through lattice architectures.
Computational Design Phase (2010s)	Simulation and optimization	Topology optimization and numerical modeling	Advanced computational tools allowed systematic design, analysis, and optimization of complex metamaterial geometries with predictable and tunable mechanical behavior.
Additive Manufacturing Era (Present)	Fabrication and scalability	High-resolution 3D printing	Additive manufacturing enabled the precise fabrication of intricate metamaterial architectures, transforming metamaterials from theoretical concepts into practical engineering solutions.
Emerging Trends (Future)	Intelligent metamaterials	Programmable and adaptive systems	Integration of smart materials, multi-stability, and active control is leading to reconfigurable and multifunctional metamaterials for advanced engineering and architectural applications.

1.3 CLASSIFICATION OF METAMATERIALS

Metamaterials can be systematically classified based on the dominant physical mechanism and functional response arising from their engineered internal architecture. Unlike conventional materials, this classification is not governed by chemical composition but by geometry, topology, and deformation behavior. Such classification aids in understanding design principles and selecting appropriate metamaterial types for specific engineering and architectural applications (Buriak et al., 2016).

Based on their primary functional characteristics, metamaterials may be broadly classified into mechanical, acoustic, electromagnetic, and multifunctional metamaterials. Within mechanical metamaterials, further classification is commonly performed by mechanical behavior, including auxetic, negative stiffness, pentamode, and programmable metamaterials. Auxetic metamaterials exhibit a negative Poisson's ratio, expanding laterally under tensile loading and offering enhanced indentation resistance. Negative stiffness metamaterials incorporate locally unstable elements that, when stabilized, provide exceptional damping and energy dissipation. Pentamode metamaterials behave as solids in shape but fluids in shear, enabling controlled stress transmission.

Additionally, phononic and acoustic metamaterials are classified based on their ability to control elastic and sound wave propagation through the formation of bandgaps, making them effective for vibration and noise mitigation. Programmable and reconfigurable metamaterials, inspired by origami and kirigami principles, represent an emerging class capable of adapting their mechanical properties in response to external stimuli.

Figure 1.3 presents a systematic classification of metamaterials based on their dominant physical mechanisms and functional responses arising from engineered internal architectures. The figure categorizes metamaterials into four major groups: mechanical, acoustic, electromagnetic, and multifunctional metamaterials. Mechanical metamaterials are shown to include auxetic structures exhibiting negative Poisson's ratio, negative stiffness metamaterials characterized by high damping and energy dissipation, and pentamode metamaterials that display fluid-like behavior under shear while maintaining solid stability. Acoustic and phononic metamaterials are illustrated

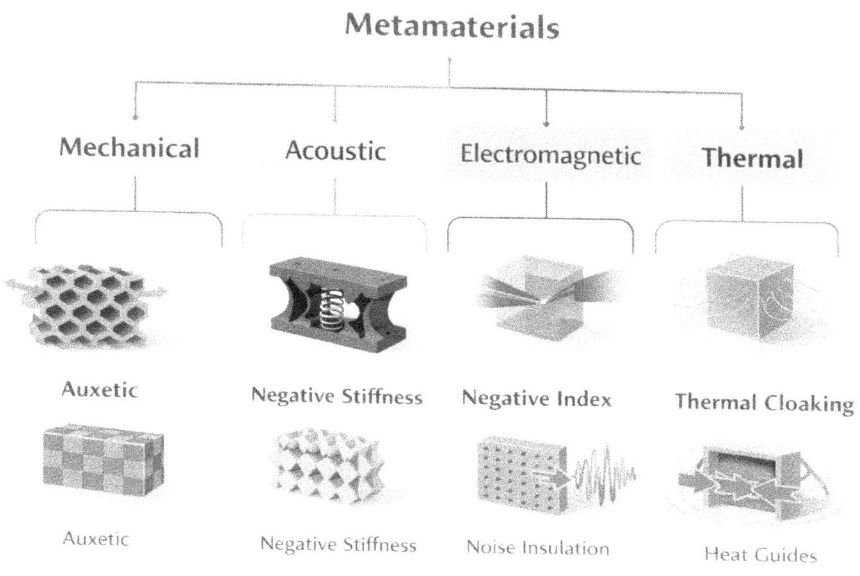

FIGURE 1.3 Classification of metamaterials.

as periodic structures capable of controlling vibration and sound propagation through the formation of mechanical bandgaps. Electromagnetic metamaterials are classified by their ability to manipulate electromagnetic waves, enabling phenomena such as negative refractive index, electromagnetic bandgaps, and cloaking effects. Multifunctional metamaterials integrate multiple physical responses within a single architected system, providing combined capabilities such as energy absorption, thermal regulation, and impact mitigation.

Mechanical metamaterials are a class of engineered materials whose mechanical properties are governed primarily by their designed internal architecture rather than by their chemical composition. By tailoring unit-cell geometry, connectivity, and topology, these materials exhibit unconventional, often counterintuitive mechanical behaviors not found in natural or conventional engineering materials.

A defining characteristic of mechanical metamaterials is their ability to achieve geometry-driven mechanical responses, such as negative Poisson's ratio (auxetic behavior), negative stiffness, extreme energy absorption, and programmable stiffness. For example, auxetic mechanical metamaterials expand laterally when stretched, resulting in enhanced indentation resistance and shear performance. Negative stiffness metamaterials incorporate locally unstable elements that, when stabilized within a structure, provide exceptional damping and vibration isolation. Pentamode mechanical metamaterials behave like solids in shape but exhibit fluid-like behavior in shear, enabling precise control of stress transmission.

Mechanical metamaterials can be designed to be lightweight, multifunctional, and tunable, making them highly attractive for advanced engineering applications. They are widely explored in aerospace and automotive structures, vibration and impact mitigation systems, robotics and soft actuators, and biomedical devices such as implants and prosthetics. Overall, mechanical metamaterials represent a shift from traditional material selection to material-by-design, where mechanical performance is programmed through structure.

1.3.1 Mechanical Metamaterials

Mechanical metamaterials are a unique class of artificially engineered materials whose mechanical performance is determined primarily by their carefully designed internal architecture rather than by the intrinsic properties or chemical composition of the constituent material. Unlike conventional engineering materials, where mechanical characteristics such as stiffness, strength, and elasticity depend mainly on material chemistry, mechanical metamaterials derive their exceptional properties from the geometry, topology, connectivity, and spatial arrangement of their repeating unit cells.

A defining characteristic of mechanical metamaterials is their ability to achieve geometry-driven mechanical responses, such as negative Poisson's ratio (auxetic behavior), negative stiffness, extreme energy absorption, and programmable stiffness. For example, auxetic mechanical metamaterials expand laterally when stretched, resulting in enhanced indentation resistance and shear performance. Negative stiffness metamaterials incorporate locally unstable elements that, when stabilized within a structure, provide exceptional damping and vibration isolation. Pentamode mechanical metamaterials behave like solids in shape but exhibit fluid-like behavior in shear, enabling precise control of stress transmission.

Mechanical metamaterials can be designed to be lightweight, multifunctional, and tunable, making them highly attractive for advanced engineering applications. They are widely explored in aerospace and automotive structures, vibration and impact mitigation systems, robotics and soft actuators, and biomedical devices such as implants and prosthetics. Overall, mechanical metamaterials represent a shift from traditional material selection to material-by-design, where mechanical performance is programmed through structure.

Figure 1.4 illustrates the major classes of mechanical metamaterials, highlighting how distinct internal architectures lead to fundamentally different mechanical responses. The figure presents representative examples of auxetic, negative stiffness, pentamode, and programmable mechanical metamaterials. Auxetic structures are shown expanding laterally under tensile loading, demonstrating a negative Poisson's ratio and improved resistance to indentation and shear. Negative stiffness metamaterials are depicted with internally unstable elements embedded within a stable framework, enabling enhanced damping and efficient energy dissipation under dynamic loading. Pentamode metamaterials are characterized by lattice-like architectures that maintain structural stability while exhibiting fluid-like behavior under shear, enabling controlled stress transmission. Programmable mechanical metamaterials, inspired by origami and reconfigurable geometries, are shown to adapt their shape and stiffness in response to external stimuli. Overall, Figure 1.4 emphasizes that the mechanical behavior of these materials is governed by geometry and topology rather than material composition, underscoring the design-driven nature of mechanical metamaterials and their potential for advanced engineering applications.

1.3.2 Acoustic Metamaterials

Acoustic metamaterials are a class of engineered materials designed to control, manipulate, and attenuate sound and elastic waves through carefully designed internal architectures. Unlike conventional acoustic materials, which rely on mass, density, or damping alone, acoustic metamaterials achieve extraordinary sound-control capabilities by exploiting periodic geometries, local resonances, and wave–structure interactions.

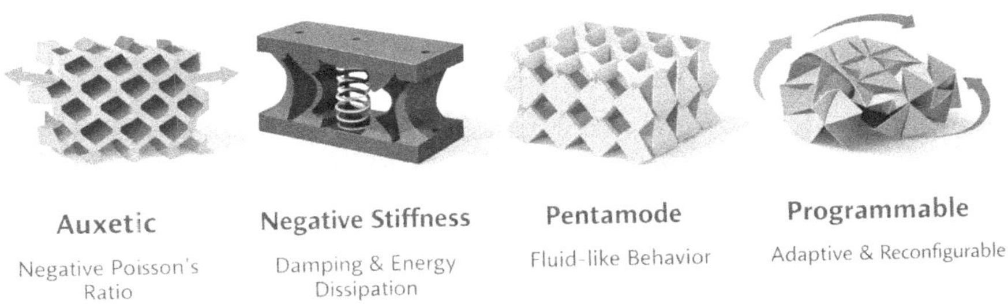

FIGURE 1.4 Mechanical metamaterials.

A defining feature of acoustic metamaterials is their ability to create acoustic or phononic bandgaps, frequency ranges in which sound or vibration propagation is prohibited. This behavior arises from periodic arrangements of unit cells that cause destructive interference or local resonance, effectively blocking or redirecting specific sound frequencies. As a result, acoustic metamaterials can achieve high sound insulation and vibration suppression even at low frequencies, where traditional materials are typically ineffective (Fok et al., 2008).

Acoustic metamaterials are widely applied in noise reduction, vibration isolation, and sound shielding. In engineering systems, they are used to protect machinery, reduce structural vibrations, and enhance acoustic comfort. In architectural applications, acoustic metamaterials enable advanced soundproofing solutions for auditoriums, buildings, and urban environments without requiring bulky or heavy materials.

Figure 1.5 illustrates the fundamental concepts and functional mechanisms of acoustic metamaterials, emphasizing their ability to control sound and vibration through engineered internal architecture. The figure highlights three representative forms: phononic structures, acoustic bandgap metamaterials, and noise insulation panels. Phononic structures are shown as periodic lattices that interact with elastic waves, enabling effective vibration and noise control by altering wave paths within the material. Acoustic bandgap metamaterials are depicted as architected unit-cell assemblies that prevent the propagation of sound waves within specific frequency ranges, thereby achieving selective wave filtering and attenuation. The noise insulation structure demonstrates how locally resonant or perforated metamaterial panels significantly reduce transmitted sound energy, even at low frequencies where conventional materials are ineffective. Overall, Figure 1.5 conveys that acoustic metamaterials achieve superior sound and vibration control not through mass or damping alone, but through geometry-driven wave manipulation, making them highly efficient, lightweight, and suitable for advanced engineering and architectural acoustic applications.

1.3.3 Electromagnetic Metamaterials

Electromagnetic metamaterials are artificially engineered materials that control and manipulate electromagnetic (EM) waves through precisely structured sub-wavelength architectures. Unlike conventional electromagnetic materials, whose properties are governed by intrinsic permittivity and permeability, electromagnetic metamaterials

FIGURE 1.5 Mechanical metamaterials.

FIGURE 1.6 Electromagnetic metamaterials.

derive their effective behavior primarily from geometry, periodicity, and spatial arrangement of their unit cells.

A defining feature of electromagnetic metamaterials is their ability to exhibit non-natural electromagnetic properties, such as negative permittivity, negative permeability, or simultaneously negative refractive index. These properties enable extraordinary phenomena, including reverse wave propagation, electromagnetic band gaps, super-lensing, and cloaking effects. By tailoring the shape and arrangement of resonant elements, electromagnetic waves can be bent, confined, filtered, or completely suppressed within specific frequency ranges.

Electromagnetic metamaterials have found wide application in antenna design, microwave and terahertz devices, sensors, imaging systems, and stealth technologies. Their compact size and tunable response enable significant miniaturization of devices while enhancing performance. Overall, electromagnetic metamaterials represent a major advancement in wave-control technology, demonstrating how engineered structures can fundamentally redefine electromagnetic behavior beyond the limits of natural materials (Walser, 2001).

Figure 1.6 illustrates the key functional mechanisms of electromagnetic metamaterials, highlighting their ability to manipulate electromagnetic waves through engineered sub-wavelength architectures. The figure presents three representative categories: negative-index metamaterials, electromagnetic bandgap structures, and cloaking metamaterials. Negative-index metamaterials are shown to bend electromagnetic waves in unconventional directions, demonstrating simultaneous negative permittivity and permeability that result in reverse refraction and unique wave-guiding behavior. Electromagnetic bandgap metamaterials are depicted as periodic structures that inhibit the propagation of electromagnetic waves within specific frequency ranges, enabling effective wave filtering, confinement, and suppression. Cloaking metamaterials are illustrated by guiding electromagnetic waves smoothly around an object, thereby minimizing scattering and rendering the object electromagnetically invisible.

1.4 DISTINCTION BETWEEN CONVENTIONAL AND MECHANICAL METAMATERIALS

Conventional materials and mechanical metamaterials differ fundamentally in the origin of their mechanical behavior. Conventional materials such as metals, polymers, and ceramics derive their properties primarily from chemical composition, atomic bonding, and microstructure. Their mechanical response stiffness, strength, ductility, and Poisson's ratio generally lie within well-established limits dictated by classical material science and continuum mechanics.

In contrast, mechanical metamaterials derive their effective properties primarily from their engineered internal architecture rather than from the base material itself. Through deliberate design of unit cells, lattice topology, and connectivity, mechanical metamaterials can exhibit non-intuitive and unconventional behaviors such as negative Poisson's ratio (auxeticity), negative stiffness, ultra-high energy absorption, and controllable wave propagation. Importantly, these properties can be tuned by modifying geometry without changing material composition.

Furthermore, while conventional materials are largely homogeneous and isotropic at the macroscopic scale, mechanical metamaterials are typically heterogeneous and architected, often displaying anisotropic and scale-dependent behavior. This distinction enables mechanical metamaterials to function as materials-by-design, offering a level of functional customization and performance optimization that is unattainable with traditional materials. As a result, mechanical metamaterials represent a paradigm shift from material selection to structural design in modern engineering.

Table 1.2 highlights the fundamental differences between conventional materials and mechanical metamaterials. While conventional materials rely on composition-driven properties with limited tunability, mechanical metamaterials achieve superior and unconventional performance through geometry-driven design, offering unprecedented flexibility and functionality for modern engineering applications.

TABLE 1.2
Distinction Between Conventional and Mechanical Metamaterials

Aspect	Conventional Materials	Mechanical Metamaterials	Description
Basis of Properties	Chemical composition and microstructure	Engineered geometry and architecture	Conventional materials depend on atomic bonding and composition, whereas metamaterials rely on designed internal structures to achieve desired behavior.
Design Approach	Material selection and processing	Architecture and unit-cell design	Performance in conventional materials is achieved by choosing materials and treatments, while metamaterials are designed through geometric tailoring.
Mechanical Behavior	Predictable and within classical limits	Unconventional and tunable	Mechanical metamaterials can exhibit behaviors such as negative Poisson's ratio and negative stiffness, which are not found in natural materials.
Property Tunability	Limited after fabrication	Highly tunable through geometry	In metamaterials, properties can be adjusted by modifying structural parameters without changing the base material.

(*Continued*)

TABLE 1.2 (Continued)
Distinction Between Conventional and Mechanical Metamaterials

Aspect	Conventional Materials	Mechanical Metamaterials	Description
Homogeneity	Generally homogeneous	Architected and heterogeneous	Conventional materials are macroscopically uniform, whereas metamaterials consist of periodic or aperiodic architectures.
Scale Dependence	Largely scale-independent	Strongly scale-dependent	Metamaterial behavior depends on unit-cell size and arrangement, allowing multiscale property control.
Manufacturing Methods	Casting, forging, machining	Additive manufacturing, micro-fabrication	Mechanical metamaterials often require advanced fabrication techniques to realize complex geometries.
Functional Integration	Single or limited functionality	Multifunctional capability	Metamaterials can integrate stiffness, damping, energy absorption, and wave control within a single material system.
Application Scope	Traditional engineering uses	Advanced and adaptive systems	Mechanical metamaterials enable innovative applications in aerospace, robotics, biomedical devices, and architecture.

1.5 HISTORICAL MILESTONES AND KEY DISCOVERIES

The development of mechanical metamaterials is rooted in a series of theoretical and experimental breakthroughs that progressively shifted material design from composition-based approaches to architecture-driven engineering. Early foundations were laid through studies on cellular solids and lattice structures, which demonstrated that geometry could significantly influence stiffness, strength, and deformation behavior. These studies provided the conceptual basis for viewing materials as structured systems rather than homogeneous continua.

A major milestone was the discovery of auxetic behavior, where certain engineered structures exhibited a negative Poisson's ratio, expanding laterally under tensile loading. This finding challenged classical elasticity theory and highlighted the potential of geometry to produce counterintuitive mechanical responses. Subsequent research introduced negative stiffness elements, revealing the possibility of achieving exceptional damping and energy dissipation when such unstable components were embedded within stable matrices.

Another key discovery involved the identification of mechanical bandgaps in periodic structures, leading to the emergence of phononic and acoustic metamaterials capable of controlling vibration and elastic wave propagation. Parallel advancements in origami- and kirigami-inspired architectures enabled reconfigurable and deployable metamaterials with programmable mechanical properties.

More recently, the integration of computational design tools, topology optimization, and additive manufacturing has transformed these concepts into practical engineering solutions. High-resolution fabrication techniques now allow precise realization of complex, hierarchical architectures, establishing mechanical metamaterials as a mature and rapidly expanding field with wide-ranging applications in engineering and architecture.

Table 1.3 summarizes the major historical milestones and key discoveries that shaped the evolution of mechanical metamaterials. The progression highlights the transition from early geometry-based concepts to advanced architected systems enabled by modern computational and manufacturing technologies, establishing mechanical metamaterials as a distinct and impactful field in engineering.

TABLE 1.3
Historical Milestones and Key Discoveries in Mechanical Metamaterials

Period	Milestone/Discovery	Key Contribution	Description
Pre-1990s	Cellular solids and lattice materials	Geometry–property relationship	Early studies showed that stiffness, strength, and deformation could be controlled through cellular and lattice architectures, forming the conceptual foundation of metamaterials.
1990s	Electromagnetic metamaterials	Geometry-driven wave control	Artificially structured materials demonstrated unusual wave behavior, introducing the idea that structure, rather than composition, governs material properties.
Early 2000s	Discovery of auxetic materials	Negative Poisson's ratio	Engineered structures exhibiting lateral expansion under tension challenged classical elasticity theory and highlighted the role of geometry in mechanical response.
Mid-2000s	Negative stiffness structures	Enhanced damping and energy absorption	Locally unstable elements embedded in stable matrices enabled exceptional vibration isolation and energy dissipation capabilities.
Late 2000s	Phononic and acoustic metamaterials	Mechanical bandgaps	Periodic structures capable of controlling elastic wave propagation led to vibration shielding and noise control applications.
2010s	Origami and kirigami metamaterials	Reconfigurable architectures	Fold-based designs introduced deployability, multi-stability, and programmable mechanical behavior.
2010s–Present	Computational design and additive manufacturing	Practical realization	Advances in numerical modeling, topology optimization, and 3D printing enabled the precise fabrication of complex metamaterial architectures for real-world applications.

1.6 SCOPE IN ENGINEERING

Mechanical metamaterials provide a transformative design framework in engineering by enabling unprecedented control over mechanical behavior through architecture. Unlike conventional materials, whose properties are fixed by composition and processing, mechanical metamaterials allow engineers to tailor stiffness, strength, deformation, energy absorption, and wave propagation simply by modifying internal geometry. This shift from material selection to structural programming significantly expands the design space available to engineers.

In aerospace and automotive engineering, mechanical metamaterials enable the development of ultra-lightweight structures with high stiffness-to-weight and strength-to-weight ratios, thereby improving fuel efficiency and structural performance. Architected crash and impact mitigation components based on metamaterial concepts offer superior energy absorption and controlled deformation, enhancing safety without additional mass. Their ability to suppress vibrations and noise also improves ride comfort and structural integrity.

Within mechanical and civil engineering, mechanical metamaterials are increasingly explored for vibration isolation, seismic protection, and infrastructure resilience. Periodic metamaterial foundations and barriers can create mechanical bandgaps that attenuate harmful vibrations, while architected energy-dissipating elements improve the durability of buildings, bridges, and industrial machinery under dynamic loading conditions (Surjadi et al., 2019).

In robotics and automation, the scope of mechanical metamaterials extends to compliant mechanisms, soft robotic systems, and adaptive grippers. Their programmable stiffness and deformability enable safe human–robot interaction, precise motion control, and multifunctional actuation. Similarly, in biomedical engineering, metamaterials enable implants, prosthetics, and orthopedic devices with tailored mechanical compatibility to biological tissues, reducing stress shielding and improving long-term performance.

Table 1.4 summarizes the wide-ranging scope of mechanical metamaterials in engineering. By enabling geometry-driven control of mechanical behavior, metamaterials support lightweight, adaptive, and high-performance solutions across aerospace, automotive, civil, biomedical, and industrial engineering applications.

TABLE 1.4
Scope of Mechanical Metamaterials in Engineering

Engineering Domain	Application Area	Key Advantages	Description
Aerospace Engineering	Lightweight structures and vibration control	High stiffness-to-weight ratio	Mechanical metamaterials enable weight reduction while maintaining structural integrity, improving fuel efficiency, and fatigue resistance.
Automotive Engineering	Crashworthiness and noise reduction	Enhanced energy absorption	Architected metamaterial components provide controlled deformation and superior impact mitigation without increasing vehicle mass.

(*Continued*)

TABLE 1.4 (Continued)
Scope of Mechanical Metamaterials in Engineering

Engineering Domain	Application Area	Key Advantages	Description
Mechanical Engineering	Vibration isolation and machinery protection	Tunable stiffness and damping	Periodic metamaterials create mechanical bandgaps that suppress harmful vibrations in rotating and heavy machinery systems.
Civil Engineering	Seismic protection and infrastructure resilience	Wave attenuation and durability	Metamaterial-based foundations and barriers reduce seismic and dynamic loads on buildings and bridges.
Robotics and Automation	Soft actuators and compliant mechanisms	Programmable deformability	Geometry-driven designs allow adaptable motion, safe interaction, and multifunctional robotic components.
Biomedical Engineering	Implants and prosthetics	Mechanical compatibility	Tunable stiffness reduces stress shielding and improves the comfort and performance of biomedical devices.
Defense and Protective Systems	Blast and impact mitigation	Extreme energy dissipation	Mechanical metamaterials enhance protection against high-strain-rate loading and ballistic impacts.
Industrial Systems	Structural optimization	Material efficiency	Metamaterials reduce material usage while maintaining or enhancing mechanical performance.

1.7 SCOPE IN ARCHITECTURE

Mechanical metamaterials introduce a new paradigm in architecture by enabling structurally efficient, adaptive, and responsive building systems through geometry-driven material design. Unlike conventional construction materials, which offer fixed mechanical behavior, metamaterials allow architects and engineers to embed functionality such as vibration control, adaptability, and energy dissipation directly into structural and façade elements.

In structural architecture, mechanical metamaterials provide innovative solutions for seismic resistance and vibration mitigation. Periodic metamaterial foundations and barriers can attenuate ground-borne vibrations and seismic waves, enhancing the resilience of buildings and critical infrastructure in earthquake-prone regions. Their ability to dissipate energy efficiently also contributes to improved structural safety under dynamic loads such as wind and traffic-induced vibrations (Montemayor et al., 2015).

In building envelopes and façades, metamaterials enable adaptive and responsive systems. Architected panels with tunable stiffness and deformation characteristics can act as smart shading devices, responsive façades, or load-bearing skins that adapt to environmental conditions. Such systems improve thermal comfort, reduce energy consumption, and support sustainable building design.

Mechanical metamaterials also support lightweight and modular construction, allowing the development of deployable, foldable, and prefabricated architectural components inspired by origami and lattice structures. These features are particularly valuable for temporary structures, emergency shelters, and rapidly deployable buildings. Additionally, acoustic metamaterials play an important role in architectural acoustics by controlling noise propagation in auditoriums, urban spaces, and transportation hubs.

Table 1.5 outlines the scope of mechanical metamaterials in architecture, highlighting their role in enhancing structural resilience, adaptability, acoustic performance, and sustainability. By embedding mechanical functionality into material architecture, metamaterials enable intelligent, efficient, and future-oriented building designs.

TABLE 1.5
Scope of Mechanical Metamaterials in Architecture

Architectural Domain	Application Area	Key Functional Benefit	Description
Structural Systems	Seismic-resistant buildings	Vibration and wave attenuation	Metamaterial-based structural elements and foundations reduce the transmission of seismic and dynamic loads, improving building resilience.
Building Foundations	Ground vibration control	Energy dissipation	Periodic metamaterial barriers attenuate ground-borne vibrations from earthquakes, traffic, and machinery.
Façades and Building Envelopes	Adaptive façades and shading	Tunable stiffness and deformation	Architected panels respond to environmental loads, improving thermal comfort and energy efficiency.
Architectural Acoustics	Noise control and sound insulation	Acoustic bandgaps	Acoustic metamaterials limit sound propagation in auditoriums, urban buildings, and transportation hubs.
Modular and Deployable Structures	Temporary and emergency buildings	Lightweight and reconfigurable design	Origami- and lattice-based metamaterials enable foldable, transportable, and rapidly deployable architectural systems.
Sustainable Architecture	Material and energy efficiency	Reduced material usage	High stiffness-to-weight ratios allow efficient use of materials while maintaining structural performance.
Urban Infrastructure	Vibration isolation for public spaces	Structural comfort	Metamaterials mitigate vibrations in bridges, walkways, and elevated structures, enhancing user comfort.
Aesthetic and Functional Design	Innovative architectural forms	Integrated functionality	Complex geometries combine structural, acoustic, and visual functions into a single architectural element.

1.8 SUMMARY

This chapter introduces the fundamental concepts of mechanical metamaterials, emphasizing a paradigm shift from composition-driven materials to architecture-driven material systems. Metamaterials are defined as artificially engineered materials whose effective properties arise primarily from their internal geometry rather than their chemical composition. The evolution of metamaterials is traced from early electromagnetic wave-manipulating structures to advanced mechanical systems enabled by computational design and additive manufacturing, highlighting how geometry became the dominant design parameter.

Metamaterials are classified into mechanical, acoustic, electromagnetic, and multifunctional metamaterials, based on their dominant physical response. Mechanical metamaterials exhibit unconventional behaviors such as negative Poisson's ratio, negative stiffness, programmable deformation, and fluid-like stress transmission, making them highly effective for energy absorption, vibration control, and adaptive structures. Acoustic metamaterials manipulate sound and elastic waves through periodic architectures and bandgaps, enabling efficient noise and vibration attenuation. Electromagnetic metamaterials control electromagnetic wave propagation, exhibiting phenomena such as negative refractive index, electromagnetic bandgaps, and cloaking effects.

A clear distinction is established between conventional materials and mechanical metamaterials: the former derive their properties from composition and microstructure, whereas the latter rely on engineered architecture to achieve tunable, non-intuitive behavior. Key historical milestones, including the discovery of auxetic behavior, negative stiffness systems, phononic bandgaps, and origami-inspired architectures, are discussed to illustrate the progression of the field. Finally, the scope of mechanical metamaterials in engineering and architecture is highlighted, demonstrating their transformative potential in lightweight structures, vibration and seismic control, adaptive façades, acoustic management, and sustainable design. Overall, the chapter establishes mechanical metamaterials as a foundational technology for next-generation, high-performance, and intelligent material systems.

REVIEW QUESTIONS

1 What are metamaterials, and how do they differ from conventional materials?
2 Why are metamaterials described as "materials-by-design"?
3 Explain the role of unit cells in determining metamaterial properties.
4 How does geometry influence the macroscopic behavior of metamaterials?
5 What is meant by architecture-driven material behavior?
6 Describe the origin of metamaterials in electromagnetic research.
7 How did the concept of metamaterials evolve into mechanical metamaterials?
8 What technological advancements accelerated the development of metamaterials?
9 Why is additive manufacturing crucial for metamaterial research?
10 Explain the transition from wave manipulation to architected mechanical systems.
11 How are metamaterials classified based on physical behavior?

12 What distinguishes mechanical metamaterials from acoustic metamaterials?
13 Define electromagnetic metamaterials with examples.
14 What are multifunctional metamaterials?
15 Why is classification based on functionality rather than composition?
16 What are mechanical metamaterials?
17 Explain auxetic behavior with reference to Poisson's ratio.
18 What are negative stiffness metamaterials and their advantages?
19 Describe pentamode metamaterials and their mechanical significance.
20 What are programmable mechanical metamaterials?
21 What are acoustic metamaterials?
22 Explain the concept of acoustic or phononic bandgaps.
23 How do acoustic metamaterials differ from traditional soundproofing materials?
24 What mechanisms enable low-frequency sound attenuation in acoustic metamaterials?
25 List applications of acoustic metamaterials in engineering and architecture.
26 What are electromagnetic metamaterials?
27 Explain the negative refractive index and its significance.
28 What are electromagnetic bandgap structures?
29 How do cloaking metamaterials work?
30 Mention key applications of electromagnetic metamaterials.
31 Compare composition-driven and geometry-driven materials.
32 Why are conventional materials limited in tunability?
33 How do mechanical metamaterials achieve unconventional behavior?
34 Explain scale dependency in metamaterials.
35 Why are metamaterials often architected and heterogeneous?
36 What role did cellular solids play in the development of metamaterials?
37 Why was the discovery of auxetic materials considered revolutionary?
38 Explain the significance of negative stiffness discoveries.
39 What are phononic metamaterials, and why are they important?
40 How did origami and kirigami concepts influence metamaterial design?
41 How do mechanical metamaterials benefit aerospace engineering?
42 What role do metamaterials play in vibration and impact mitigation?
43 Explain applications of mechanical metamaterials in robotics.
44 How are metamaterials used in biomedical engineering?
45 Why are metamaterials suitable for multifunctional engineering systems?
46 How do metamaterials contribute to seismic-resistant architecture?
47 Explain the role of metamaterials in adaptive building façades.
48 How are acoustic metamaterials applied in architectural acoustics?
49 What advantages do metamaterials offer in modular and deployable structures?
50 How do mechanical metamaterials support sustainable architectural design?

2 Design Principles and Fabrication Techniques

2.1 INTRODUCTION

Mechanical metamaterials represent a paradigm shift in material engineering, where mechanical performance is encoded through geometry rather than composition. The successful development of such materials relies on two interdependent aspects: rational design principles that define targeted mechanical behavior and advanced fabrication techniques capable of realizing complex architected geometries. This chapter presents the fundamental design philosophies governing mechanical metamaterials and reviews the fabrication technologies that enable their practical implementation.

2.2 DESIGN PRINCIPLES OF MECHANICAL METAMATERIALS

The design of mechanical metamaterials is governed by the principle that mechanical behavior is dictated primarily by geometry and architecture rather than material composition. Unlike conventional materials, where properties are intrinsic, mechanical metamaterials are deliberately engineered systems in which structure functions as the primary design variable. The key design principles are outlined below.

2.2.1 Unit Cell-Based Design

Unit cell–based design is the foundational principle in the development of mechanical metamaterials. A unit cell represents the smallest repeating structural element whose geometry, orientation, and connectivity govern the effective macroscopic behavior of the material. By periodically arranging this unit cell in one, two, or three dimensions, a bulk metamaterial is formed whose mechanical properties emerge from the collective response of the architecture rather than from the intrinsic properties of the base material.

The geometry of the unit cell plays a critical role in determining deformation mechanisms such as bending, rotation, buckling, or stretching. For instance, re-entrant and chiral unit cells can induce auxetic behavior, while lattice-based unit cells enable precise control over the stiffness-to-weight ratio and energy absorption. Variations in strut thickness, angle, length, and node connectivity allow for the systematic tuning of properties such as the elastic modulus, Poisson's ratio, and load-bearing capacity.

Unit cell–based design also enables scalability and predictability in mechanical performance. Because the global behavior is derived from a well-defined repeating element, analytical and numerical homogenization techniques can be applied to estimate effective material properties. This makes unit cell–based design particularly

DOI: 10.1201/9781003783671-2

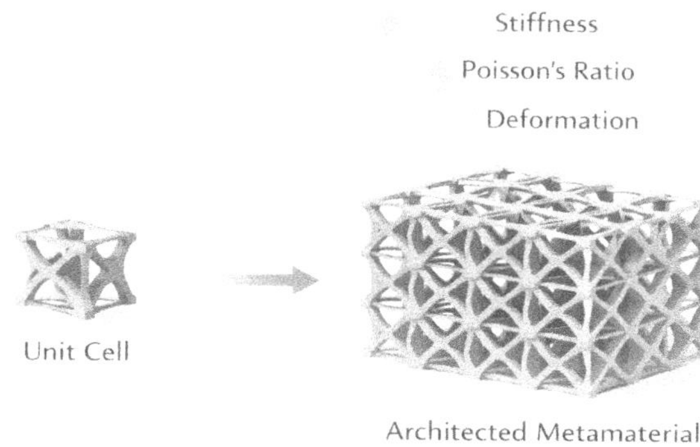

FIGURE 2.1 Unit cell–based design of mechanical metamaterials.

attractive for computational optimization and additive manufacturing, where complex architectures can be efficiently generated and fabricated (Akhmetshin et al. 2023).

Figure 2.1 illustrates the principle of unit cell–based design, which forms the foundation of mechanical metamaterials. The figure conceptually shows a single unit cell as the smallest repeatable structural element, characterized by its geometry, dimensions, and connectivity. When this unit cell is periodically replicated in one, two, or three dimensions, it forms a bulk architected metamaterial structure. The macroscopic mechanical behavior, such as stiffness, Poisson's ratio, deformation mode, and energy absorption emerges from the collective response of these repeating unit cells rather than from the intrinsic properties of the base material. By modifying unit-cell parameters such as strut lengths, angles, thicknesses, and nodal arrangements, designers can systematically tailor global mechanical performance.

2.2.2 Geometry and Topology Control

Geometry and topology control is a core design principle of mechanical metamaterials, wherein the shape, arrangement, and connectivity of structural elements dictate the overall mechanical response. Unlike conventional materials, whose behavior is primarily governed by intrinsic material properties, mechanical metamaterials exploit geometric features such as strut orientations, node configurations, and void distributions to activate specific deformation mechanisms.

Geometric control influences how a structure deforms under load. By designing unit cells that promote bending, rotation, buckling, or hinging, metamaterials can exhibit large elastic strains, auxetic behavior, or enhanced energy absorption. For example, bending-dominated geometries provide high deformability and damping, whereas stretching-dominated geometries yield higher stiffness and load-bearing capacity.

Topological control refers to the connectivity and load path network within the metamaterial. The manner in which elements are interconnected determines stress distribution, stability, and failure modes. Topology optimization techniques are often employed to identify optimal material layouts that achieve desired performance

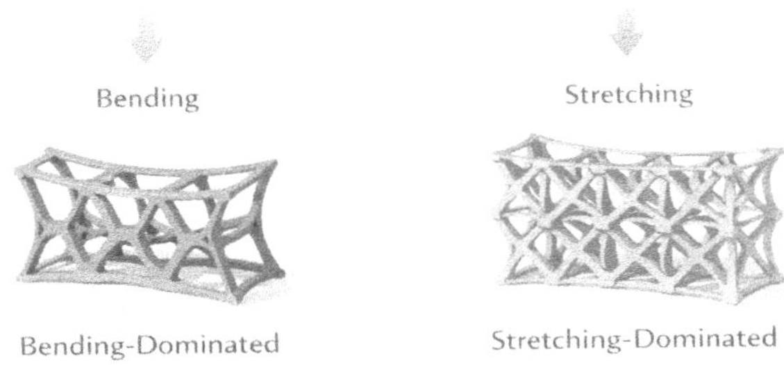

FIGURE 2.2 Geometry and topology control of mechanical metamaterials.

targets such as minimum weight, maximum stiffness, or controlled instability. Through topological tailoring, mechanical metamaterials can achieve unconventional responses including negative stiffness, mechanical bandgaps, and anisotropic behavior.

Together, geometry and topology control enable designers to program mechanical behavior at the architectural level, allowing for the precise tuning of stiffness, strength, deformation, and wave propagation. This principle is fundamental to achieving the extraordinary and application-specific performance that defines mechanical metamaterials.

Figure 2.2 illustrates how geometry and topology govern the deformation mechanisms and mechanical performance of mechanical metamaterials. The figure compares two architected lattices designed with different topological arrangements: a bending-dominated structure and a stretching-dominated structure. In the bending-dominated configuration, the geometry promotes flexural deformation of struts and rotational motion at the joints when subjected to external loading. This results in high compliance, large recoverable deformation, and enhanced energy dissipation, making such architectures suitable for impact mitigation and damping applications. In contrast, the stretching-dominated configuration features a topology where loads are primarily carried through the axial stretching of structural members. This topology provides higher stiffness, strength, and load-bearing capacity with relatively lower deformation. The figure clearly demonstrates that by controlling geometric features such as strut orientation and node connectivity, designers can selectively activate desired deformation modes.

2.2.3 Periodic Architectures

Periodic architectures in mechanical metamaterials are formed by the uniform repetition of a single unit cell throughout the structure in one, two, or three dimensions. This strict translational symmetry leads to homogeneous and predictable effective properties, making periodic designs the most widely studied and implemented class of architected metamaterials.

From a mechanical standpoint, periodicity ensures that deformation and load transfer are evenly distributed across the structure. As a result, properties such as elastic modulus, the Poisson's ratio, stiffness-to-weight ratio, and deformation mode can be accurately predicted using analytical homogenization or numerical simulations. This predictability is particularly advantageous for structural design, vibration isolation, and wave-control applications.

Periodic architectures also play a critical role in mechanical and acoustic bandgap formation. The regular spacing of structural elements causes constructive and destructive interference of elastic waves, enabling the selective attenuation of vibration or sound within specific frequency ranges. This makes periodic metamaterials highly effective for vibration suppression, noise control, and phononic applications (Sharma & Hiremath 2022).

In terms of fabrication, periodic architectures are well-suited to additive manufacturing, as their repetitive nature simplifies design automation, scalability, and quality control. However, while periodic designs offer uniform performance, they may be less effective in handling localized damage or highly non-uniform loading conditions in comparison with graded or aperiodic architectures.

Figure 2.3 illustrates the concept of periodic architectures, where a single unit cell is uniformly repeated throughout the material to form an architected mechanical metamaterial. The figure highlights a representative unit cell and its periodic replication in three dimensions, resulting in a lattice with translational symmetry. This uniform repetition ensures homogeneous and predictable mechanical properties, such as stiffness, Poisson's ratio, and deformation behavior across the entire structure. The regular geometric arrangement also enables controlled wave propagation, as periodicity can give rise to mechanical or acoustic bandgaps that attenuate vibrations within specific frequency ranges. By maintaining identical geometry and connectivity at every repetition, periodic architectures simplify modeling, homogenization, and fabrication—making them particularly suitable for additive manufacturing (AM) and scalable structural applications.

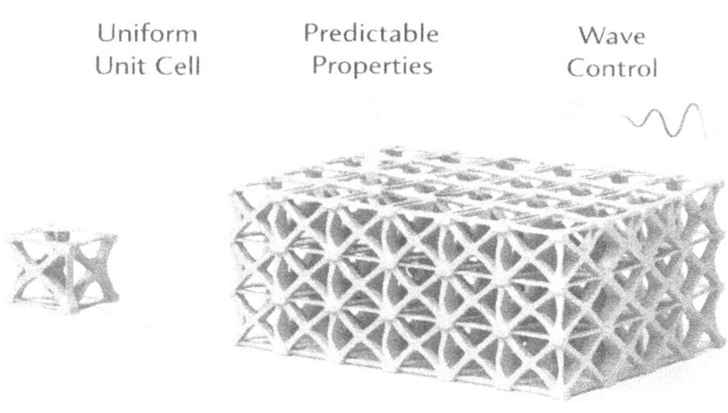

FIGURE 2.3 Periodic architectures of mechanical metamaterials.

2.2.4 Aperiodic Architectures

Aperiodic architectures in mechanical metamaterials are characterized by the absence of strict translational symmetry, meaning that unit cells are arranged in a non-repeating or quasi-periodic manner. Unlike periodic architectures, where identical unit cells are uniformly repeated, aperiodic designs deliberately introduce structural irregularity to achieve enhanced and more robust mechanical performance.

From a mechanical perspective, aperiodicity disrupts uniform load paths and wave propagation patterns. This disruption leads to distributed stress fields, reduced stress concentrations, and improved damage tolerance under complex or localized loading conditions. As a result, aperiodic mechanical metamaterials are particularly effective for impact mitigation, crash protection, and broadband vibration attenuation, where periodic structures may exhibit localized failure or narrow-band responses.

Aperiodic architectures also play a crucial role in wave manipulation, as the lack of periodicity suppresses coherent wave interference and reduces resonance effects. This enables broader frequency attenuation in comparison with periodic metamaterials, making aperiodic designs attractive for vibration and noise control applications requiring wide operational bandwidths.

Designing aperiodic metamaterials often relies on computational optimization techniques, such as stochastic design, graded randomness, or topology optimization, to balance mechanical efficiency with robustness. Although the modeling and fabrication of aperiodic architectures are more complex than those of their periodic counterparts, advances in AM have made their practical realization increasingly feasible (Imediegwu et al. 2023).

Figure 2.4 illustrates the concept of aperiodic architectures, where mechanical metamaterials are constructed using irregular or non-repeating arrangements of unit cells. Unlike periodic architectures, the unit cells in this structure do not follow strict translational symmetry, resulting in a heterogeneous lattice. The figure highlights how such irregularity leads to distributed stress fields, as indicated by the spread of deformation and stress across the structure rather than localization at a single point. This redistribution enhances energy dissipation and reduces stress

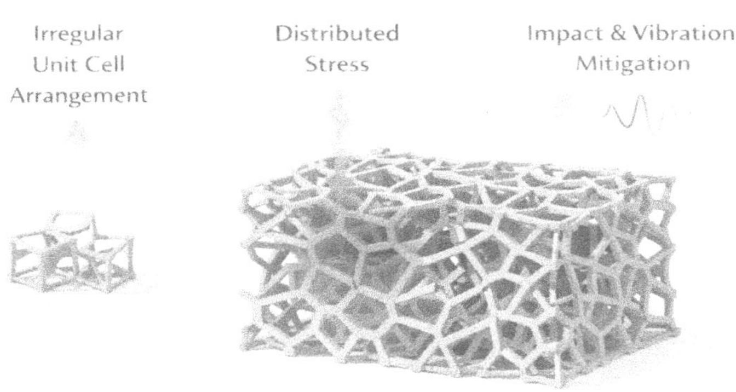

FIGURE 2.4 Aperiodic architectures of mechanical metamaterials.

concentration, making aperiodic architectures particularly effective for impact and vibration mitigation. Additionally, the disruption of regular load paths and wave propagation suppresses resonance effects, enabling broader frequency attenuation in comparison with periodic structures.

2.2.5 Graded Architectures

Graded architectures, also known as functionally graded metamaterials, are characterized by a systematic spatial variation in unit-cell geometry, density, or topology across the structure. Unlike periodic architectures with uniform properties or aperiodic architectures with irregular arrangements, graded metamaterials are intentionally designed to exhibit smooth transitions in the mechanical response from one region to another.

The primary advantage of graded architectures lies in their ability to tailor local mechanical properties such as stiffness, strength, and energy absorption according to functional requirements. By gradually varying parameters like strut thicknesses, cell sizes, orientations, or connectivities, graded metamaterials achieve controlled stress redistribution, reduced stress concentrations, and improved structural integrity under complex loading conditions.

Graded architectures are particularly effective in impact mitigation and load-transfer applications, where a progressive change in stiffness allows for the efficient absorption and dissipation of energy. In biomedical engineering, stiffness-graded metamaterials are used in implants to match the mechanical properties of bone and surrounding tissues, minimizing stress shielding. In structural and architectural applications, graded metamaterials enable lightweight components with enhanced durability and damage tolerance (Zhang et al. 2021).

Figure 2.5 illustrates the concept of graded (functionally graded) architectures in mechanical metamaterials, where mechanical properties vary smoothly across the structure through the systematic modification of unit-cell geometry. The figure shows a continuous transition from a region with relatively compliant, low-density unit cells to a region with stiffer, denser unit cells. This gradual variation enables progressive stiffness and location-specific mechanical behavior, allowing for efficient load

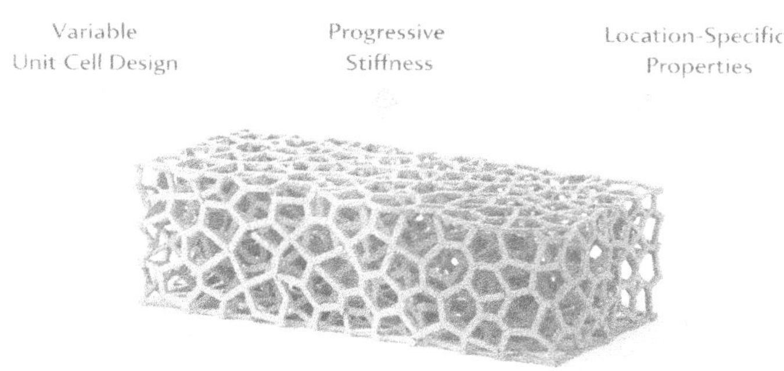

FIGURE 2.5 Graded architectures of mechanical metamaterials.

transfer and stress redistribution under external loading. By avoiding abrupt changes in geometry, graded architectures reduce stress concentrations and enhance energy absorption and damage tolerance. Such architectures are particularly advantageous in impact mitigation, biomedical implants, and lightweight structural components, where different regions must satisfy distinct mechanical requirements.

2.2.6 Multistability and Programmability

Multistability and programmability enable mechanical metamaterials to switch among multiple stable configurations, thereby altering their mechanical response on demand. In multistable metamaterials, the architecture is designed to possess two or more distinct equilibrium states (bistable or multistable) separated by energy barriers. Transitions between these states can be triggered by mechanical loading, thermal stimuli, magnetic or electric actuation, or embedded smart materials.

From a mechanics standpoint, multistability is commonly achieved by exploiting geometric nonlinearity, elastic buckling, snap-through instabilities, and compliant mechanisms. These effects allow large, reversible shape changes without material failure. Each stable state can exhibit a different stiffness, Poisson's ratio, load path, or deformation mode, effectively enabling property switching within the same material system.

Programmable metamaterials extend this concept by allowing the sequence, direction, or extent of deformation to be predefined or reconfigured. Programming can be achieved through unit-cell geometry, spatial arrangement of bistable elements, or external control fields. Origami- and kirigami-inspired architectures are prominent examples, where folding patterns encode mechanical functionality such as deployability, shape morphing, and adaptive stiffness.

The practical significance of multistability and programmability lies in their ability to create adaptive and reconfigurable systems. Applications include soft robotics, deployable aerospace structures, vibration isolation systems with tunable response, and adaptive architectural components. Overall, this design principle transforms mechanical metamaterials from passive load-bearing materials into active, intelligent systems capable of responding dynamically to changing operational requirements (He et al. 2024).

Figure 2.6 illustrates the principle of multistability and programmability in mechanical metamaterials. The figure shows a single architected structure capable of existing in multiple stable configurations, labeled as Stable State 1 and Stable State 2. These distinct equilibrium states are separated by an energy barrier and can be accessed through external stimuli such as mechanical loading, actuation, or controlled deformation. The transition between states is enabled by geometric nonlinearity and elastic instabilities, such as snap through or buckling, which allow large, reversible shape changes without permanent damage.

The figure further highlights programmability, where the sequence and final configuration of deformation can be predefined or reconfigured. Each stable state exhibits different mechanical characteristics, such as stiffness, shape, or load-transfer paths, enabling property switching within the same material system. This capability transforms mechanical metamaterials from passive structures into adaptive and reconfigurable systems, suitable for applications in soft robotics, deployable structures, vibration control, and adaptive architectural components.

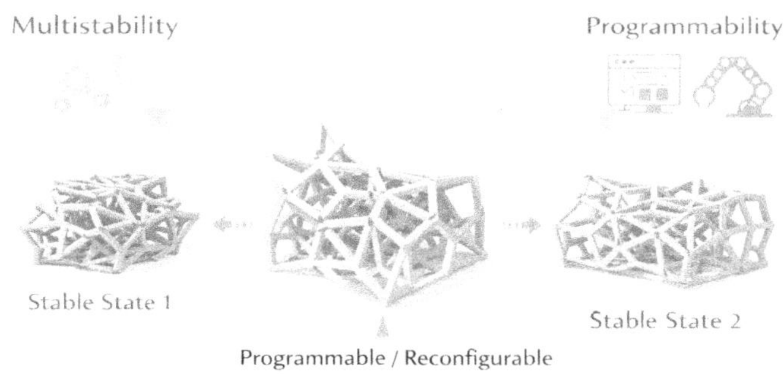

FIGURE 2.6 Multistability and programmability in mechanical metamaterials.

2.2.7 Hierarchical and Multiscale Design

Hierarchical and multiscale design is a key principle in mechanical metamaterials wherein architectural features are intentionally introduced across multiple length scales from micro- and meso-scale unit cells to macroscale structural organization. Rather than relying on a single characteristic size, hierarchical metamaterials combine nested or layered architectures so that mechanical performance emerges from interactions across scales.

At the microscale, fine structural features govern local deformation mechanisms such as bending, buckling, or rotation, enabling high compliance, energy dissipation, or auxetic behavior. At the mesoscale, these micro-architected units are assembled into lattices or cellular networks that control stiffness, strength, and load transfer. At the macroscale, the overall geometry defines global structural performance, stability, and boundary interactions. This multilevel organization enables mechanical metamaterials to achieve combinations of properties, such as high strength with low density or high toughness with recoverable deformation, that are difficult to realize with single-scale designs.

Hierarchical design also enhances damage tolerance and robustness. When damage initiates at one scale, other scales can redistribute loads and delay catastrophic failure. This concept is strongly inspired by natural materials such as bone, wood, and nacre, where hierarchical organization provides exceptional mechanical efficiency. In engineering applications, hierarchical mechanical metamaterials are particularly valuable for lightweight load-bearing structures, impact-resistant components, and energy-absorbing systems (Song et al. 2024).

Figure 2.7 illustrates the concept of hierarchical and multiscale design, where mechanical metamaterials are structured across multiple length scales to achieve superior and multifunctional mechanical performance. The figure highlights three distinct levels of organization: the microscale, mesoscale, and macroscale. At the microscale, fine structural features govern local deformation mechanisms such as bending, buckling, and rotation, enabling enhanced compliance and energy dissipation. These micro-architected elements are assembled at the mesoscale into lattice or cellular networks that control stiffness, strength, and load transfer. At the macroscale, the overall geometry defines global structural behavior, stability, and boundary interactions.

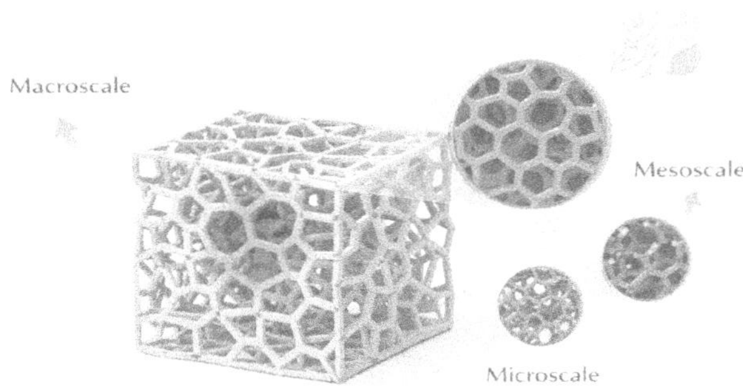

FIGURE 2.7 Hierarchical and multiscale design of mechanical metamaterials.

This hierarchical organization allows mechanical metamaterials to combine lightweight characteristics with high strength, improved damage tolerance, and efficient energy absorption. The figure also reflects bio-inspired design principles, where multiscale architectures found in natural materials are replicated to achieve optimized mechanical efficiency.

2.2.8 Computational Design and Optimization

Computational design and optimization play a central role in the development of mechanical metamaterials, enabling systematic and performance-driven architectural design. Unlike conventional materials, where properties are intrinsic, mechanical metamaterials require a precise geometric configuration to achieve targeted responses, such as auxeticity, negative stiffness, tunable bandgaps, or high energy absorption. Computational methods provide the necessary framework to predict, optimize, and validate these complex behaviors prior to fabrication (Bonfanti et al. 2024).

Finite Element Analysis (FEA): Finite element modeling is widely used to analyze stress distribution, deformation modes, buckling behavior, and the dynamic response of metamaterial unit cells and periodic assemblies. FEA allows the evaluation of nonlinear effects, instability-driven mechanisms, and multistable transitions under realistic loading conditions. It also supports homogenization techniques for estimating effective elastic properties.

Topology Optimization: Topology optimization determines the optimal material distribution within a given design domain to achieve specific objectives, such as maximum stiffness, minimum weight, or controlled deformation. In mechanical metamaterials, topology optimization is frequently used to generate novel unit-cell geometries that exhibit unconventional mechanical responses while maintaining structural efficiency.

Parametric and Generative Design: Parametric modeling enables the systematic variation of geometric parameters, such as strut thickness, cell angle, and connectivity. Generative design approaches automate geometry creation based on performance constraints, expanding the design space beyond traditional, intuition-based methods.

Inverse Design Methods: Inverse design begins with desired macroscopic properties (e.g., a target Poisson's ratio or bandgap frequency) and computes the corresponding optimal architecture. This approach is increasingly supported by machine learning algorithms and data-driven modeling techniques.

Multiscale Modeling: Multiscale computational frameworks link microscale unit-cell behavior to macroscale structural performance. Such models ensure the accurate prediction of effective properties and structural reliability in hierarchical metamaterials.

Figure 2.8 presents the integrated computational workflow used in the design and optimization of mechanical metamaterials. At the center of the figure is an architected lattice structure representing the candidate metamaterial geometry. Surrounding it are the major computational stages that collectively define performance-driven design.

The process begins with the specification of structural objectives and constraints, such as target stiffness, weight reduction, energy absorption capacity, or a prescribed Poisson's ratio. Based on these requirements, parametric and generative design tools are employed to create candidate unit-cell geometries by systematically varying geometric parameters or automatically generating optimized configurations within defined boundaries.

These candidate designs are evaluated using FEA to assess stress distribution, deformation modes, instability behavior, and the dynamic response under realistic loading conditions. The results guide refinement through topology optimization, where material distribution within the design domain is iteratively adjusted to maximize structural efficiency or achieve targeted mechanical behavior.

The figure also highlights multiscale modeling, which links microscale unit-cell behavior to mesoscale lattice performance and macroscale structural response. This ensures that local deformation mechanisms translate accurately into effective bulk properties.

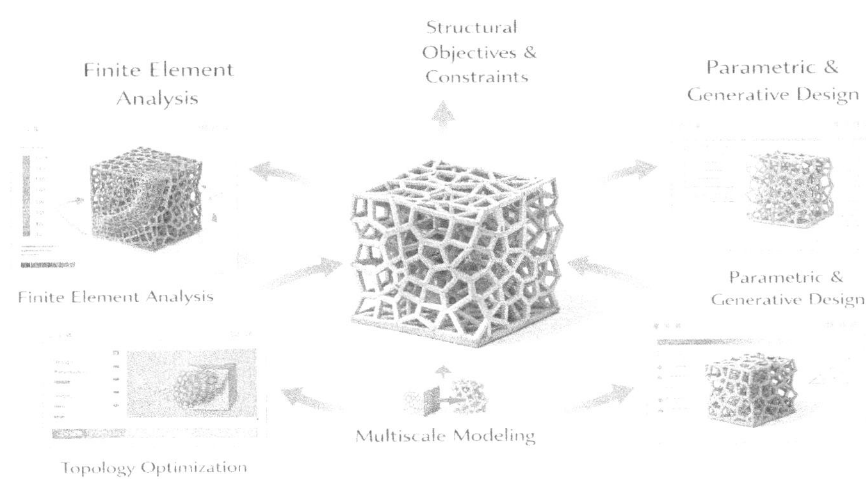

FIGURE 2.8 Computational design and optimization of mechanical metamaterials.

2.3 FABRICATION TECHNIQUES FOR MECHANICAL METAMATERIALS

The practical realization of mechanical metamaterials requires fabrication techniques capable of producing complex, high-precision, and often multiscale architectures. Since mechanical performance is governed by geometry, manufacturing accuracy directly influences stiffness, strength, stability, and wave-control behavior. Fabrication methods must therefore ensure geometric fidelity, scalability, and material compatibility.

2.3.1 Polymer Additive Manufacturing (3D Printing)

AM, commonly known as 3D printing, is the most transformative fabrication technique for mechanical metamaterials. Since metamaterial performance depends critically on internal architecture, AM provides the geometric freedom required to fabricate complex lattice, cellular, and hierarchical structures with high precision.

Unlike conventional subtractive manufacturing, AM builds components layer-by-layer directly from digital models, enabling the realization of intricate unit-cell geometries, internal voids, graded structures, and multiscale architectures that are otherwise impossible to fabricate (Khosravani et al. 2025).

2.3.1.1 Fused Deposition Modeling (FDM)

FDM is a material extrusion-based AM technique widely used for fabricating polymer-based mechanical metamaterials. The process operates on the principle of **layer-by-layer deposition of a thermoplastic material**, enabling the direct conversion of a digital three-dimensional model into a physical component. Due to its accessibility, cost-effectiveness, and ability to produce complex geometries, FDM is frequently employed in the prototyping and experimental validation of lattice and architected metamaterial structures Cuan-Urquizo and Guerra Silva, 2023).

2.3.1.1.1 Working Principle

The working mechanism of FDM is based on the controlled melting and extrusion of a thermoplastic filament. A continuous filament of material, commonly polylactic acid (PLA), acrylonitrile butadiene styrene (ABS), or polyethylene terephthalate glycol (PETG) is fed into a heated extrusion head. The material is heated above its glass transition or melting temperature until it reaches a semi-molten state. The softened material is then extruded through a fine nozzle and deposited onto a build platform in accordance with predefined toolpaths generated by slicing software. As the material is deposited, it cools and solidifies, bonding with the underlying layer. By sequentially depositing layers in the vertical (Z) direction while controlling horizontal (X–Y) motion, a three-dimensional structure is gradually constructed.

2.3.1.1.2 Process Sequence

The FDM process begins with the creation of a digital model using computer-aided design (CAD) software. The model is subsequently processed using slicing software, which divides the geometry into thin layers and generates machine-readable instructions (G-code). During printing, the filament is continuously supplied from a spool to the extruder assembly, where it is driven into the heated nozzle. The molten material is extruded and precisely deposited onto the build platform following the programmed path. Once each layer is completed, the build platform or print head moves incrementally to accommodate the next layer. After fabrication, support structures if used are removed, and minor post-processing may be performed to improve surface finish.

2.3.1.1.3 Main Components

An FDM system consists of several essential components:

- **Filament Spool**: Supplies thermoplastic feedstock.
- **Extruder Assembly**: Controls filament feeding and extrusion rate.
- **Hot End and Nozzle**: Melt and deposit the material with controlled precision.
- **Build Platform (Print Bed)**: Provides a stable surface for layer deposition, often heated to enhance adhesion.
- **Motion Control System**: Guides movement along the X, Y, and Z axes.
- **Control Unit and Software**: Regulate temperature, motion, and process parameters.
- **Cooling System**: Assists in rapid solidification and interlayer bonding.

Figure 2.9 illustrates the working principle and main components of the FDM process used for fabricating mechanical metamaterials. The figure shows a typical FDM system where a thermoplastic filament is supplied from a filament spool and fed into the extruder assembly. The extruder pushes the filament into the hot

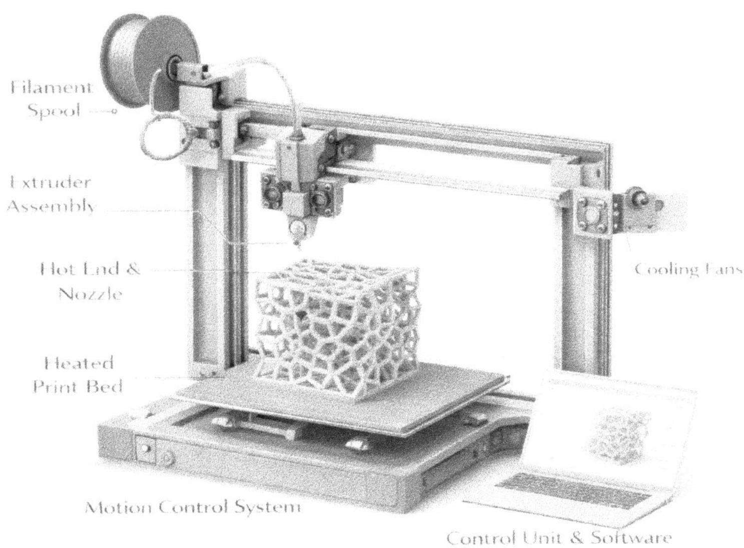

FIGURE 2.9 Schematic of FDM system and printing process.

end and nozzle, where it is heated above its melting temperature and extruded in a semi-molten state.

The molten material is deposited layer-by-layer onto a heated print bed, forming the lattice-based metamaterial structure. The motion control system governs the precise movement of the print head along the X–Y axes and controls vertical displacement along the Z-axis to enable successive layer deposition. Cooling fans assist in rapid solidification and improve interlayer bonding. The entire process is regulated by a control unit and software, which interprets slicing instructions (G-code) derived from the original CAD model.

The figure highlights the layer-wise construction mechanism that enables fabrication of complex lattice and cellular geometries commonly used in mechanical metamaterials. It also emphasizes the dependence of structural accuracy and mechanical performance on controlled temperature, extrusion rate, and motion precision.

Table 2.1 presents commonly used materials in the FDM process along with their key properties and applications. Material selection in FDM-based mechanical metamaterials depends on the required stiffness, strength, ductility, thermal resistance, and functional performance. For example, PLA is preferred for rapid prototyping and educational models due to its ease of printing, whereas nylon and polycarbonate are

TABLE 2.1
Materials Used in FDM Process

Material	Key Mechanical/ Thermal Properties	Advantages in FDM	Typical Applications (Including Metamaterials)
PLA (polylactic acid)	Tensile strength: ~50–70 MPa; Low warping; Tg ~60°C	Easy to print; Biodegradable; Good surface finish	Prototyping of lattice structures, auxetic models, educational metamaterials
ABS (acrylonitrile butadiene styrene)	Tensile strength: ~30–45 MPa; Tg ~105°C; Impact resistant	Higher toughness than PLA; Heat resistant	Functional lattice components, impact studies, structural prototypes
PETG (polyethylene terephthalate glycol)	Tensile strength: ~45–55 MPa; Good ductility; Tg ~80°C	Good layer adhesion; Chemical resistance	Durable metamaterial lattices, graded structures
TPU (thermoplastic polyurethane)	High elongation (>300%); Flexible; Shore hardness variable	Flexible and elastic; Good impact resistance	Soft metamaterials, compliant mechanisms, energy-absorbing structures
Nylon (polyamide)	Tensile strength: ~50–75 MPa; High fatigue resistance; Tough	Strong and durable; Good wear resistance	Load-bearing lattices, mechanical joints, structural metamaterials
PC (polycarbonate)	Tensile strength: ~60–70 MPa; Tg ~147°C; High impact strength	High temperature resistance; Strong	High-performance structural metamaterials

(Continued)

TABLE 2.1 (Continued)
Materials Used in FDM Process

Material	Key Mechanical/ Thermal Properties	Advantages in FDM	Typical Applications (Including Metamaterials)
Carbon fiber-reinforced PLA/ ABS	High stiffness; Improved strength-to-weight ratio	Lightweight and rigid; Reduced deformation	Aerospace-inspired lattice metamaterials
Glass fiber-reinforced Filaments	Improved modulus; Enhanced dimensional stability	Higher stiffness than base polymer	Structural lattice components
PEEK (polyether ether ketone)	Tensile strength: ~90–100 MPa; Tg ~143°C; High thermal resistance	Excellent mechanical and thermal performance	High-performance aerospace and biomedical metamaterials
HIPS (high impact polystyrene)	Moderate strength; Soluble in limonene	Good support material; Lightweight	Support structures, lightweight prototypes

selected for load-bearing or high-performance metamaterial structures. Flexible materials such as TPU are suitable for compliant and energy-absorbing metamaterials. Reinforced composites provide enhanced stiffness and are particularly useful for lightweight structural applications.

2.3.1.2 Stereo Lithography (SLA)

Stereolithography (SLA) is a vat photopolymerization-based AM technique that fabricates three-dimensional components through the selective curing of liquid photopolymer resin using ultraviolet (UV) radiation. Introduced as one of the earliest 3D printing technologies, SLA is recognized for its high dimensional accuracy, fine feature resolution, and superior surface finish. These characteristics make it particularly suitable for the fabrication of intricate lattice and micro-architected structures used in mechanical metamaterials (Xiao et al. 2021).

2.3.1.2.1 Working Principle

The fundamental working principle of SLA is based on photopolymerization, a chemical process in which liquid resin solidifies when exposed to a specific wavelength of light. The system consists of a resin-filled vat and a build platform. A UV laser beam or digital light projection system selectively scans the surface of the resin according to the cross-sectional geometry of the first layer. The exposed regions undergo polymerization and solidify, forming a thin solid layer.

After completing one layer, the build platform moves vertically either downward into the resin or upward depending on the printer configuration, allowing fresh liquid resin to flow over the previously cured layer. The laser then scans the next cross-sectional pattern. This layer-by-layer process continues until the entire three-dimensional structure is completed.

2.3.1.2.2 Process Sequence

The SLA fabrication process begins with the creation of a digital model using CAD software. The model is converted into a set of thin slices using slicing software, typically with layer thicknesses ranging from 25 to 100 micrometers. During printing, the UV light source cures each layer sequentially according to the sliced geometry. Once printing is complete, the part is removed from the build platform and washed to remove uncured resin. A post-curing process under controlled UV exposure is often required to improve mechanical properties and ensure complete polymerization.

2.3.1.2.3 Main Components

An SLA system comprises several essential components:

- **Resin Vat**: Contains the liquid photopolymer resin.
- **UV Laser or Projector**: Provides the energy required for selective curing.
- **Build Platform**: Supports the structure during fabrication.
- **Recoating Mechanism**: Ensures uniform distribution of resin between layers.
- **Control System and Software**: Regulate scanning patterns, exposure time, and layer thickness.

Figure 2.10 illustrates the working principle and essential components of the SLA process used for fabricating high-resolution mechanical metamaterials. The figure shows a vat photopolymerization system in which a UV laser is directed by galvanometer mirrors to selectively cure a liquid photopolymer resin contained within a resin vat. The laser scans the predefined cross-sectional geometry of each layer, initiating photopolymerization and solidifying the resin at precise locations.

The partially fabricated metamaterial lattice is attached to a build platform, which moves incrementally along the vertical axis after each layer is cured. This controlled vertical displacement allows fresh liquid resin to flow beneath the previously solidified layer, enabling sequential layer formation. The illustration also highlights the liquid photopolymer resin, which remains uncured until exposed to UV radiation.

After printing, the component undergoes a post-curing process, typically in a UV curing chamber, to complete polymerization and enhance mechanical strength and dimensional stability. The control unit and software manage scanning patterns, exposure time, layer thickness, and overall process parameters.

Table 2.2 summarizes commonly used photopolymer materials in SLA-based AM along with their key properties and applications. Material selection for SLA-fabricated mechanical metamaterials depends on the required stiffness, toughness, flexibility, and thermal resistance. Standard resins are suitable for high-precision prototypes, while tough and flexible resins are preferred for energy-absorbing and compliant metamaterials. Rigid and ceramic-filled resins are selected for stiffness-dominated structures, whereas high-temperature resins are used in thermally demanding environments. Biocompatible resins enable biomedical metamaterial applications, such as scaffolds and implants.

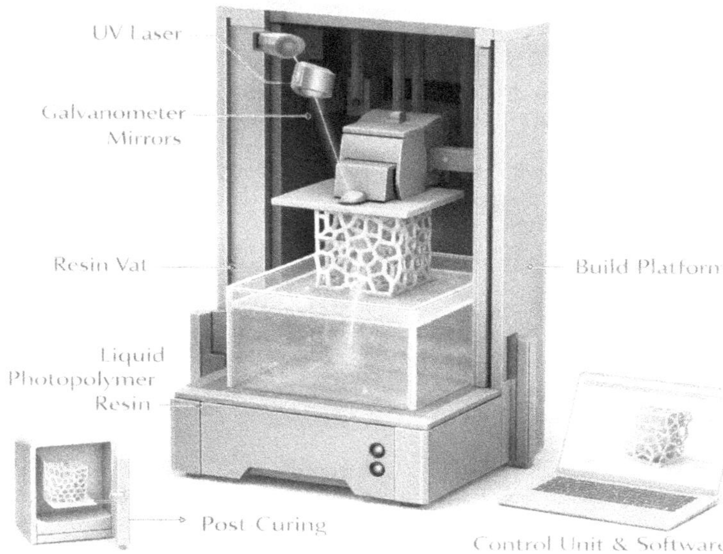

FIGURE 2.10 Schematic of SLA system with key components.

TABLE 2.2
Materials Used in SLA Process

Material Type	Key Mechanical/ Thermal Properties	Advantages in SLA	Typical Applications (Including Metamaterials)
Standard Photopolymer Resin	Moderate tensile strength (~40–65 MPa); Brittle behavior; Tg ~50–60°C	High resolution; Smooth surface finish	High-precision lattice prototypes, auxetic structures, visualization models
Tough Resin	Improved impact resistance; Higher elongation than standard resin	Enhanced durability; Reduced brittleness	Functional lattice components, snap-fit metamaterials
Rigid Resin	High modulus (~2–4 GPa); Low deformation	Dimensional stability; Stiff structures	Load-bearing lattice prototypes, structural metamaterials
Flexible Resin	High elongation (>150%); Shore hardness 50–80A	Elastic behavior; Energy absorption	Soft metamaterials, compliant mechanisms
Durable Resin	Low friction; Moderate flexibility; Good wear resistance	Fatigue resistance; Smooth finish	Moving metamaterial joints, mechanical linkages

(Continued)

TABLE 2.2 (Continued)
Materials Used in SLA Process

Material Type	Key Mechanical/ Thermal Properties	Advantages in SLA	Typical Applications (Including Metamaterials)
High-Temperature Resin	Heat deflection temperature >200°C (post-cured); High stiffness	Thermal resistance	Aerospace-inspired metamaterials, heat-exposed components
Castable Resin	Burns out cleanly; Low ash residue	High detail reproduction	Mold fabrication for metamaterial casting
Dental/ Biocompatible Resin	Biocompatible; Moderate strength	Medical compliance	Biomedical lattice implants, scaffolds
Ceramic-Filled Resin	Increased stiffness; Improved thermal stability	Enhanced rigidity	High-modulus micro-lattice structures
Composite (Glass-Filled) Resin	Higher modulus than standard resin; Reduced shrinkage	Improved structural rigidity	Graded and topology-optimized metamaterials

2.3.1.3 Selective Laser Sintering (SLS)

SLS is a powder bed fusion-based AM technique that fabricates three-dimensional components through the selective sintering of powdered material using a high-energy laser beam. Unlike extrusion- or resin-based systems, SLS utilizes thermoplastic or composite powders that are fused layer-by-layer without the need for support structures. Due to its ability to produce mechanically robust and complex internal geometries, SLS is widely employed in the fabrication of mechanical metamaterials, particularly lattice and cellular architectures requiring structural integrity (Yuan et al., 2019).

2.3.1.3.1 Working Principle

The working principle of SLS is based on the localized thermal fusion of powder particles. A thin layer of powder material is evenly spread across the build platform using a recoating blade or roller. A computer-controlled laser selectively scans the cross-sectional geometry of the part, raising the temperature of the powder above its sintering point. The heated particles partially melt and fuse together, forming a solid layer.

Once a layer is completed, the build platform lowers by a predetermined layer thickness, and a new layer of powder is spread across the surface. The laser then sinters the next cross-section. This process is repeated until the entire three-dimensional structure is formed. The unsintered powder surrounding the part acts as a natural support medium, enabling the fabrication of complex overhangs and intricate lattice structures without additional support materials.

2.3.1.3.2 Process Sequence

The SLS process begins with the preparation of a CAD model, which is converted into layered instructions using slicing software. During printing, the powder bed is preheated to a temperature slightly below the melting point of the material to reduce thermal gradients and residual stresses. The laser selectively sinters each layer according to the digital pattern. After completion, the entire powder bed is allowed to cool gradually to prevent warping. The finished component is then removed from the surrounding powder, which can often be recycled for future builds.

2.3.1.3.3 Main Components

An SLS system consists of the following essential components:

- **Powder Bed/Build Chamber**: Contains the powder material and fabricated part.
- **Laser Source (CO_2 or Fiber Laser)**: Provides the energy required for sintering.
- **Recoating Mechanism**: Spreads thin, uniform powder layers.
- **Build Platform**: Moves vertically to accommodate successive layers.
- **Heating System**: Maintains controlled thermal conditions within the chamber.
- **Control Unit and Software**: Regulate laser scanning, temperature, and layer thickness.

2.3.1.3.4 Materials Used

Common materials for SLS include nylon (PA12, PA11), glass-filled nylon, carbon fiber reinforced nylon, thermoplastic elastomers (TPE), and certain metal powders (in related powder bed fusion systems). Nylon-based powders are particularly favored for mechanical metamaterials due to their balance of strength, toughness, and fatigue resistance.

Figure 2.11 illustrates the working principle and key components of the SLS process used for fabricating mechanical metamaterials. The figure shows a powder bed

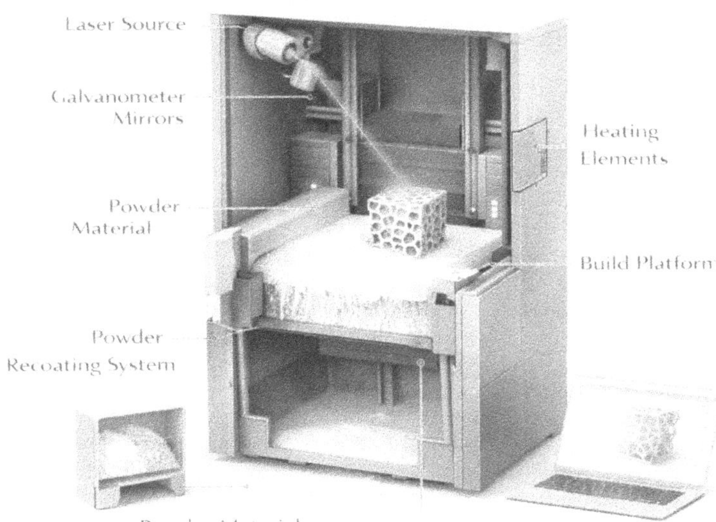

FIGURE 2.11 Selective Laser Sintering (SLS) process and system components.

fusion system in which a high-energy laser source is directed by galvanometer mirrors to selectively sinter powdered material according to the cross-sectional geometry of the designed structure.

A thin layer of powder material is uniformly spread across the build platform using a powder recoating system. The laser scans the designated areas, locally heating the powder particles above their sintering temperature, causing them to fuse and form a solid layer. Surrounding unsintered powder remains loose and acts as a natural support for overhangs and intricate lattice features. After completion of one layer, the build platform moves downward by one layer thickness, and a fresh layer of powder is spread. This layer-by-layer sintering continues until the complete three-dimensional metamaterial structure is formed.

The figure also highlights the heating elements, which maintain the powder bed at a temperature slightly below the melting point to reduce thermal gradients and residual stresses. Upon completion of printing, the part is allowed to cool gradually within the powder bed before being removed and cleaned.

Table 2.3 summarizes commonly used materials in the SLS process along with their key properties and typical applications. Nylon-based materials such as PA12 and PA11 dominate SLS due to their balanced strength, toughness, and process stability. Reinforced variants (glass- or carbon-filled) provide enhanced stiffness for structural metamaterials. Elastomeric powders enable the fabrication of flexible and

TABLE 2.3
Materials Used in Selective Laser Sintering (SLS)

Material	Key Mechanical/ Thermal Properties	Advantages in SLS	Typical Applications (Including Metamaterials)
PA12 (Nylon 12)	Tensile strength: ~45–55 MPa; Good fatigue resistance; Moderate flexibility; Melting point ~178°C	Excellent dimensional stability; Good surface finish; Recyclable powder	Load-bearing lattice metamaterials, impact-resistant structures
PA11 (Nylon 11)	Tensile strength: ~45–50 MPa; Higher ductility than PA12; Bio-based	High toughness; Improved elongation	Energy-absorbing and flexible metamaterial architectures
Glass-filled nylon (PA12-GF)	Increased modulus (~2–3 GPa); Reduced elongation	Higher stiffness; Improved thermal stability	Stiff structural lattices, graded metamaterials
Carbon fiber-reinforced nylon	High stiffness-to-weight ratio; Enhanced strength	Lightweight; Reduced deformation	Aerospace-inspired lattice metamaterials
Thermoplastic elastomers (TPE/ TPU)	High elasticity; Elongation >200%; Soft	Flexible; Energy absorption	Soft mechanical metamaterials, compliant mechanisms

(Continued)

TABLE 2.3 (Continued)
Materials Used in Selective Laser Sintering (SLS)

Material	Key Mechanical/ Thermal Properties	Advantages in SLS	Typical Applications (Including Metamaterials)
Polypropylene (PP)	Low density; Good chemical resistance; Moderate flexibility	Lightweight; Chemical stability	Lightweight graded structures, fluid-exposed components
Polystyrene (PS)	Moderate stiffness; Brittle	Fine detail capability	Prototyping of complex lattice geometries
Alumide (aluminum-filled nylon)	Improved thermal conductivity; Increased stiffness	Enhanced rigidity; Metallic appearance	Structural prototypes, semi-rigid metamaterials
PEEK (in high-temp SLS systems)	Tensile strength: ~90–100 MPa; High thermal resistance; Tg ~143°C	High-performance polymer; Excellent mechanical strength	Aerospace, biomedical lattice implants
Metal powders (in related SLM/DMLS processes)	High strength; Excellent load-bearing capability	Structural durability; High temperature resistance	Metallic metamaterials, high-strength lattice components

energy-absorbing metamaterial systems. High-performance polymers such as PEEK are used in advanced aerospace and biomedical applications.

Material selection in SLS-based mechanical metamaterials depends on the required balance between stiffness, toughness, thermal stability, and density. Nylon-based powders remain the most widely used for architected lattice structures due to their reliable sintering behavior and mechanical robustness.

Table 2.4 compares FDM, SLA, and SLS additive manufacturing processes in terms of material type, mechanical properties, accuracy, and suitability for mechanical metamaterials.

TABLE 2.4
Comparative Analysis of FDM, SLA, and SLS Materials

Parameter	FDM (Material Extrusion)	SLA (Vat Photopolymerization)	SLS (Powder Bed Fusion – Polymer)
Primary Material Form	Thermoplastic filament	Liquid photopolymer resin	Polymer powder
Common Materials	PLA, ABS, PETG, Nylon, TPU, PC, CF-reinforced filaments	Standard resin, Tough resin, Rigid resin, Flexible resin, High-temp resin	PA12, PA11, Glass-filled Nylon, Carbon-filled Nylon, TPE, PP
Typical Tensile Strength	30–75 MPa (depends on material)	40–65 MPa (resin dependent)	45–55 MPa (PA12 typical)

(Continued)

TABLE 2.4 (Continued)
Comparative Analysis of FDM, SLA, and SLS Materials

Parameter	FDM (Material Extrusion)	SLA (Vat Photopolymerization)	SLS (Powder Bed Fusion – Polymer)
Elastic Modulus Range	1–3 GPa (PLA/ABS); higher with reinforcement	1.5–4 GPa (rigid resins)	1.5–2.5 GPa (nylon); higher with fillers
Elongation at Break	Low–moderate (PLA low, TPU high)	Generally low (except flexible resins)	Moderate (higher ductility than SLA)
Thermal Resistance	Moderate (up to ~150°C for PC)	Moderate (high-temp resin >200°C HDT post-cure)	Good (~170–180°C melting for nylon)
Surface Finish	Moderate; visible layer lines	Excellent; smooth finish	Moderate; slightly rough powder texture
Dimensional Accuracy	Moderate	High	High–Moderate
Mechanical Isotropy	Anisotropic (weaker interlayer bonding)	Nearly isotropic	More isotropic than FDM
Support Structures	Required for overhangs	Required	Not required (powder acts as support)
Best Suited For	Rapid prototyping, low-cost lattice models	High-resolution micro-architectures	Functional, load-bearing lattice metamaterials
Metamaterial Applications	Auxetic models, graded structures, impact studies	Micro-lattices, precision unit cells, biomedical prototypes	Structural lattices, impact-resistant metamaterials

2.3.2 Metal Additive Manufacturing

Metal additive manufacturing (metal AM) refers to a group of advanced fabrication technologies that produce fully dense metallic components through the layer-by-layer consolidation of metal powders using high-energy heat sources. Among the most widely adopted metal AM processes are selective laser melting (SLM), direct metal laser sintering (DMLS), and electron beam melting (EBM). These processes belong to the powder bed fusion family and are extensively used for fabricating complex, high-performance mechanical metamaterials intended for load-bearing and structural applications (Suhas et al., 2025).

2.3.2.1 Working Principle

The fundamental principle of metal AM is based on the localized melting and solidification of metal powder particles. A thin layer of fine metal powder is first spread uniformly across a build platform inside a controlled chamber. A high-energy heat source, either a focused laser beam (SLM/DMLS) or an electron beam (EBM), selectively scans the cross-sectional geometry of the component as defined by digital slicing data. The energy input melts the powder locally, forming a dense metallic layer upon solidification.

After one layer is completed, the build platform lowers by a predetermined thickness (typically 20–60 micrometers), and a new layer of powder is recoated. The scanning and melting sequence is repeated layer-by-layer until the entire three-dimensional structure is built. The surrounding unmelted powder acts as a natural support medium during fabrication. Upon completion, the part is allowed to cool, followed by its removal from the powder bed and necessary post-processing.

2.3.2.2 Selective Laser Melting (SLM)

SLM is an advanced powder bed fusion-based metal AM process in which a high-power laser is used to fully melt and consolidate fine metal powders into dense, three-dimensional components. Unlike conventional manufacturing techniques that rely on machining or casting, SLM builds parts layer-by-layer directly from a digital model, enabling the fabrication of complex geometries that are otherwise difficult or impossible to produce. In the field of mechanical metamaterials, SLM plays a crucial role in realizing high-strength metallic lattice architectures with optimized mechanical performance.

2.3.2.2.1 Working Principle

The SLM process operates on the principle of localized laser-induced melting and rapid solidification. A thin layer of metal powder typically between 20 and 60 micrometers thick is evenly spread across a build platform inside a controlled atmosphere chamber filled with an inert gas (such as argon or nitrogen). A high-energy fiber laser selectively scans the powder surface according to the cross-sectional geometry defined by the sliced CAD model.

The laser energy is sufficient to completely melt the powder particles in the scanned regions, forming a molten pool. Upon cooling, the molten metal solidifies, creating a dense and metallurgically bonded layer. After completing one layer, the build platform lowers by one layer thickness, and a new powder layer is applied. The process is repeated layer-by-layer until the entire component is formed. The surrounding unmelted powder acts as a support medium during fabrication.

2.3.2.2.2 Materials Used

SLM is compatible with a wide range of engineering metals and alloys, including:

- Titanium alloys (e.g., Ti-6Al-4V)
- Stainless steels (e.g., 316L)
- Aluminum alloys (e.g., AlSi10Mg)
- Nickel-based superalloys (e.g., Inconel 718)
- Cobalt-chromium alloys

These materials are selected for their strength-to-weight ratio, corrosion resistance, and high-temperature performance.

2.3.2.2.3 Microstructural Characteristics

Due to the rapid heating and cooling cycles inherent in SLM, parts typically exhibit fine microstructures and high mechanical strength. However, the process may also introduce residual stresses and anisotropy depending on scanning strategy and build orientation. Post-processing treatments such as stress-relief annealing, hot isostatic pressing (HIP), and surface finishing are often required to optimize mechanical properties and dimensional accuracy.

Figure 2.12 illustrates the working principle and principal components of the SLM process used for fabricating dense metallic mechanical metamaterials. The system operates within a controlled build chamber filled with inert gas (typically argon or nitrogen) to prevent oxidation during melting.

A thin layer of metal powder is spread uniformly across the build platform by the powder recoating system. A high-energy laser source, directed by galvanometer mirrors and focused through a beam focus lens, selectively scans the powder bed according to the cross-sectional geometry derived from the CAD model. The laser energy fully melts the powder particles in the scanned region, forming a molten pool that solidifies upon cooling to create a dense metallic layer.

After the completion of one layer, the build platform lowers by a predefined layer thickness, and a new layer of powder is applied. This layer-by-layer melting process continues until the entire three-dimensional metallic lattice or structural component is formed. Heating elements help maintain controlled thermal conditions within the chamber, thereby reducing thermal gradients and minimizing residual stresses.

The figure emphasizes the powder bed fusion mechanism that enables the fabrication of complex, high-strength lattice architectures with near-full density. This capability makes SLM particularly suitable for load-bearing mechanical metamaterials in aerospace, biomedical, and structural engineering applications.

Table 2.5 presents commonly used metallic materials compatible with the SLM process, along with their key mechanical and thermal properties and applications.

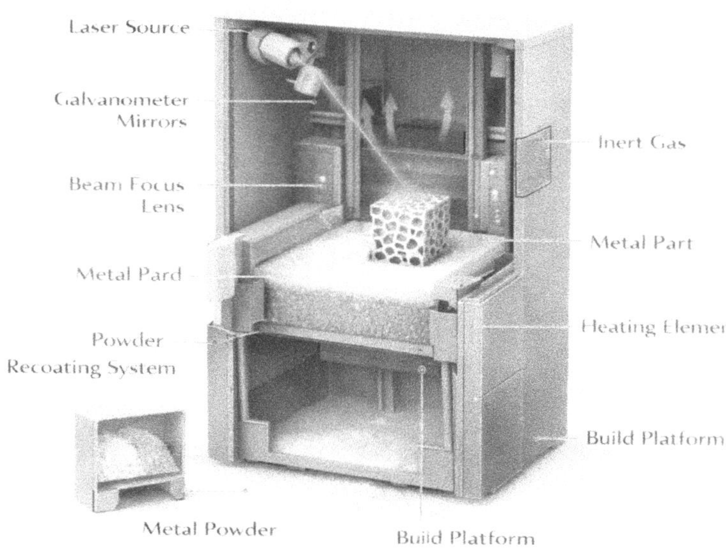

FIGURE 2.12 Selective laser melting (SLM) process and system components.

TABLE 2.5
Materials Used in Selective Laser Melting (SLM)

Material	Key Mechanical/ Thermal Properties	Advantages in SLM	Typical Applications (Including Metamaterials)
Ti-6Al-4V (titanium alloy)	Tensile strength: ~900–1100 MPa; Low density (~4.43 g/cm^3); Excellent corrosion resistance	High strength-to-weight ratio; Biocompatible	Aerospace lattice structures, biomedical implants, graded metamaterials
316L stainless steel	Tensile strength: ~500–700 MPa; Good ductility; Excellent corrosion resistance	Good weldability; Structural durability	Load-bearing lattice metamaterials, marine and chemical components
AlSi10Mg (aluminum alloy)	Tensile strength: ~350–450 MPa; Low density (~2.7 g/cm^3); Good thermal conductivity	Lightweight; Good fatigue resistance	Lightweight aerospace metamaterials, heat-dissipating lattice structures
Inconel 718 (nickel superalloy)	Tensile strength: ~1000–1300 MPa; High temperature resistance (>700°C)	Excellent creep and oxidation resistance	High-temperature lattice metamaterials, turbine components
Co-Cr (cobalt-chromium alloy)	Tensile strength: ~900–1200 MPa; High wear resistance	Biocompatible; High hardness	Orthopedic implants, dental lattice structures
Maraging steel	Tensile strength: ~1000–2000 MPa (after heat treatment); High toughness	Exceptional strength; Heat treatable	High-strength structural metamaterials, tooling lattices
Tool steel (H13)	High hardness; wear resistant; good thermal stability	Suitable for high-stress environments	Impact-resistant metamaterial structures
Copper alloys (CuCrZr)	High thermal conductivity; Moderate strength	Excellent heat dissipation	Thermal management lattice metamaterials
Nickel alloy 625	Tensile strength: ~800–1000 MPa; Corrosion resistant	Stable at high temperature	Aerospace and marine lattice structures
Aluminum 7075 (limited SLM systems)	High strength (~500 MPa); Lightweight	High strength-to-weight ratio	Lightweight structural metamaterials

Titanium alloys and aluminum alloys are widely used for lightweight structural metamaterials due to their favorable strength-to-weight ratios. Stainless steels and maraging steels are preferred for load-bearing and high-strength lattice structures. Nickel-based superalloys such as Inconel 718 are employed in high-temperature environments, while cobalt-chromium alloys are selected for biomedical applications. Copper alloys are particularly suitable for thermal management metamaterials due to their high conductivity.

Material selection in SLM-based mechanical metamaterials depends on the required strength, density, fatigue resistance, thermal stability, and application environment. The ability of SLM to fabricate dense metallic lattices enables the realization of high-performance architected materials for advanced engineering systems.

2.3.2.3 Direct Metal Laser Sintering (DMLS)

DMLS is a powder bed fusion-based metal AM process in which a high-energy laser selectively fuses fine metal powder particles to fabricate fully dense, three-dimensional metallic components. Although historically described as a "sintering" process, modern DMLS systems typically achieve near-complete melting of powder particles, producing parts with mechanical properties comparable to wrought or cast materials. DMLS is widely used for fabricating complex, high-performance components, including metallic mechanical metamaterials with optimized lattice architectures (Humnabad et al. 2022).

2.3.2.3.1 Working Principle

The DMLS process begins with the deposition of a thin, uniform layer of metal powder over a build platform inside a sealed chamber filled with an inert gas, typically argon or nitrogen. The inert atmosphere prevents oxidation and ensures metallurgical integrity during processing. A high-powered fiber laser, guided by galvanometer mirrors, scans the powder bed according to the cross-sectional geometry derived from the sliced CAD model.

The laser energy locally heats the powder particles to temperatures sufficient to cause fusion. In modern systems, this typically results in near-full melting rather than partial sintering. The fused region solidifies rapidly upon cooling, forming a dense metallic layer metallurgically bonded to the underlying layer. After one layer is completed, the build platform lowers by a small increment, usually between 20 and 60 micrometers and a new layer of powder is spread by the recoating mechanism. The process is repeated layer-by-layer until the complete part is formed.

2.3.2.3.2 Process Environment and Thermal Control

Thermal management plays a critical role in DMLS. The powder bed is often preheated to reduce thermal gradients and minimize residual stresses. However, due to the rapid heating and cooling cycles, residual stresses and distortion may still occur. Post-processing operations such as stress-relief heat treatment, support removal, machining, and surface finishing are commonly required to achieve final dimensional accuracy and optimal mechanical performance.

2.3.2.3.3 Materials Used

DMLS is compatible with a broad range of engineering metals, including stainless steels (e.g., 316L), titanium alloys (Ti-6Al-4V), aluminum alloys (AlSi10Mg), cobalt-chromium alloys, maraging steels, and nickel-based superalloys (e.g., Inconel 718). These materials are selected for their high strength, corrosion resistance, fatigue performance, and thermal stability.

Figure 2.13 illustrates the operational principle and key components of the DMLS system used for fabricating dense metallic components and lattice-based mechanical metamaterials. The system operates within a sealed build chamber maintained under an inert gas atmosphere (typically argon or nitrogen) to prevent oxidation during high-temperature processing.

A thin layer of metal powder is uniformly distributed across the build platform by the powder recoating system. A high-energy laser source, directed through galvanometer mirrors and focused via a beam focusing lens, selectively scans the powder bed according to the sliced CAD geometry. The laser energy fuses the powder particles locally, forming a solid metallic layer.

After the completion of one layer, the build platform lowers incrementally, and a fresh layer of powder is applied. This layer-by-layer fusion process continues until the final three-dimensional metallic structure is produced. Heating elements assist in maintaining controlled thermal conditions, reducing thermal gradients, and minimizing residual stresses. Surrounding unsintered powder supports the structure during fabrication and is later removed during post-processing.

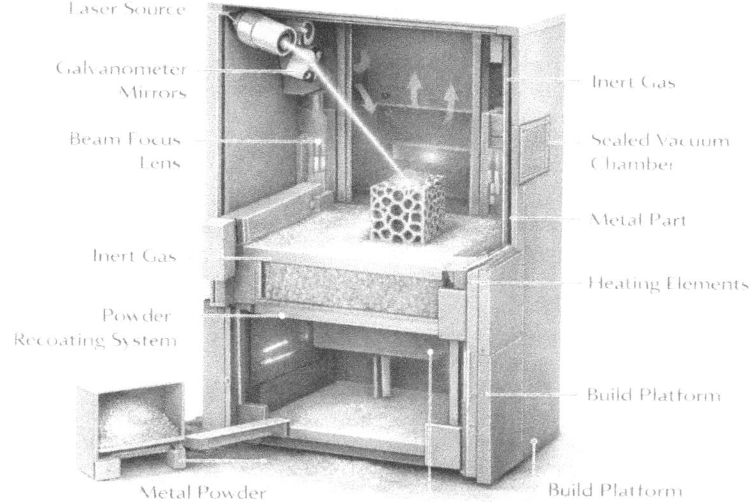

FIGURE 2.13 Direct metal laser sintering (DMLS) process and system components.

The figure highlights the powder bed fusion mechanism and the controlled thermal environment that enables fabrication of intricate, high-strength metallic metamaterial architectures suitable for structural and load-bearing applications.

Table 2.6 presents the metallic materials commonly processed by the DMLS technique, along with their primary mechanical and thermal characteristics and applications. Stainless steels and titanium alloys are widely used for structural and biomedical metamaterials due to their strength, corrosion resistance, and durability. Aluminum alloys are selected for lightweight lattice systems, while nickel-based superalloys are preferred for high-temperature applications. Maraging and tool steels are employed where extremely high strength and wear resistance are required. Copper alloys are

TABLE 2.6
Materials Used in Direct Metal Laser Sintering (DMLS)

Material	Key Mechanical/ Thermal Properties	Advantages in DMLS	Typical Applications (Including Metamaterials)
316L Stainless Steel	Tensile strength: ~500–700 MPa; Good ductility; Excellent corrosion resistance	Good weldability; Stable processing	Load-bearing lattice metamaterials, chemical and marine components
17-4 PH Stainless Steel	Tensile strength: ~900–1200 MPa (after heat treatment); High hardness	Heat treatable; High strength	Structural metamaterials, tooling lattices
Ti-6Al-4V (Titanium Alloy)	Tensile strength: ~900–1100 MPa; Low density (4.43 g/cm^3); Biocompatible	High strength-to-weight ratio; Corrosion resistant	Aerospace lattice structures, biomedical implants
AlSi10Mg (Aluminum Alloy)	Tensile strength: ~350–450 MPa; Low density (2.7 g/cm^3); Good thermal conductivity	Lightweight; Good fatigue resistance	Lightweight metamaterials, heat-exchanging lattice systems
Inconel 718 (Nickel Superalloy)	Tensile strength: ~1000–1300 MPa; Excellent high-temperature strength (>700°C)	Creep and oxidation resistant	High-temperature lattice metamaterials, turbine components
Inconel 625	Tensile strength: ~800–1000 MPa; High corrosion resistance	Stable at elevated temperatures	Aerospace and marine metamaterial structures
Cobalt-Chromium (Co-Cr)	Tensile strength: ~900–1200 MPa; High wear resistance	Biocompatible; High hardness	Orthopedic and dental lattice implants
Maraging Steel (MS1)	Tensile strength: ~1000–2000 MPa (after aging); High toughness	Exceptional strength; Heat treatable	High-strength metamaterials, tooling structures
Tool Steel (H13)	High hardness; good thermal stability; wear resistant	Suitable for impact-resistant parts	Industrial metamaterial components
Copper Alloys (CuCrZr)	High thermal conductivity; Moderate strength	Excellent heat dissipation	Thermal management lattice metamaterials

particularly suitable for thermal management metamaterials due to their high thermal conductivity.

Material selection in DMLS-based mechanical metamaterials depends on the required mechanical strength, density, fatigue performance, corrosion resistance, and service environment. The capability of DMLS to fabricate dense metallic lattice structures enables the development of high-performance architected materials for advanced engineering applications.

2.3.2.4 Electron Beam Melting (EBM)

EBM is an advanced powder bed fusion-based metal AM process that utilizes a high-energy electron beam to selectively melt metal powder in a layer-by-layer manner. Unlike laser-based systems such as SLM and DMLS, EBM operates under a high-vacuum environment and employs an electron beam as the energy source. The process is particularly well-suited for reactive metals such as titanium alloys and for applications requiring reduced residual stress and high structural integrity. In the field of mechanical metamaterials, EBM enables the fabrication of robust, high-performance metallic lattice structures with complex geometries (Korner, 2016).

2.3.2.4.1 Working Principle

The EBM process begins with the deposition of a thin layer of metal powder across a build platform inside a vacuum chamber. The vacuum environment is essential to prevent electron scattering and oxidation during processing. A high-energy electron beam, generated by an electron gun and directed by electromagnetic coils, scans the powder bed according to the cross-sectional geometry obtained from the sliced CAD model.

Before full melting occurs, the powder layer is typically preheated using a defocused electron beam. This preheating step lightly sinters the powder particles, stabilizing the powder bed and minimizing thermal distortion during subsequent melting. The electron beam then selectively melts the required regions, forming a dense metallic layer. After the completion of one layer, the build platform lowers by a specified layer thickness, and a new powder layer is spread. This sequence continues until the entire three-dimensional component is produced.

2.3.2.4.2 Thermal Characteristics

EBM operates at relatively high build temperatures compared to laser-based systems. The elevated processing temperature reduces thermal gradients and significantly lowers residual stresses within the fabricated component. As a result, parts produced by EBM generally exhibit reduced distortion and improved dimensional stability. The slower cooling rates also influence microstructural characteristics, often leading to favorable mechanical properties.

2.3.2.4.3 Materials Used

EBM is commonly used for processing the below:

- Titanium alloys (e.g., Ti-6Al-4V)
- Cobalt-chromium alloys
- Nickel-based superalloys
- Certain high-performance stainless steels

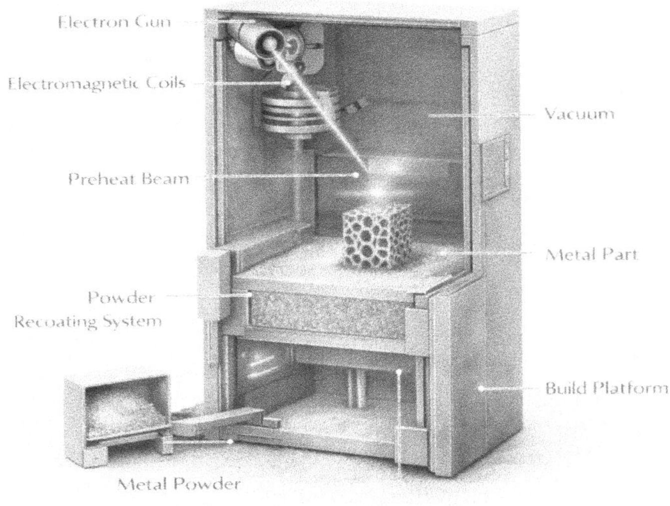

FIGURE 2.14 Schematic of EBM system.

Titanium alloys are particularly suitable due to their compatibility with vacuum environments and excellent strength-to-weight ratio.

Figure 2.14 illustrates the operational principle and major components of the EBM system used for fabricating dense metallic components and lattice-based mechanical metamaterials. The system operates within a vacuum chamber, which is essential to prevent electron scattering and oxidation during high-energy processing.

At the top of the system, an electron gun generates a focused beam of high-velocity electrons. The beam is directed and controlled by electromagnetic coils, which precisely guide it across the powder bed. Before full melting, a preheating beam lightly sinters the powder layer to stabilize the powder bed and reduce thermal distortion. The focused electron beam then selectively melts the required regions of the metal powder according to the sliced CAD geometry, forming a dense metallic layer.

A powder recoating system spreads a uniform thin layer of metal powder over the build platform after each layer is completed. The build platform lowers incrementally to accommodate successive layers, enabling layer-by-layer fabrication of the metallic structure. Due to the elevated build temperature maintained within the chamber, thermal gradients are minimized, which significantly reduces residual stresses in comparison with laser-based systems.

Table 2.7 summarizes the metallic materials commonly processed using the EBM technique, along with their primary mechanical properties and applications. Titanium alloys dominate EBM applications due to their compatibility with vacuum processing, an excellent strength-to-weight ratio, and reduced residual stress formation. Cobalt-chromium alloys are widely used in biomedical applications due to their wear resistance and biocompatibility. Nickel-based superalloys such as Inconel grades are preferred for high-temperature structural applications.

TABLE 2.7
Materials Used in Electron Beam Melting (EBM)

Material	Key Mechanical/ Thermal Properties	Advantages in EBM	Typical Applications (Including Metamaterials)
Ti-6Al-4V (Titanium Alloy)	Tensile strength: ~900–1100 MPa; Low density (4.43 g/cm³); Excellent corrosion resistance; Good fatigue strength	Ideal for vacuum processing; Low residual stress; Biocompatible	Aerospace lattice metamaterials, biomedical implants, graded porous structures
Ti-6Al-4V ELI (Extra Low Interstitial)	Improved fracture toughness; Enhanced biocompatibility	Superior ductility; Reduced impurities	Orthopedic lattice implants, bone-mimicking metamaterials
Co-Cr (Cobalt-Chromium Alloy)	Tensile strength: ~900–1200 MPa; High wear and corrosion resistance	High hardness; Stable at elevated temperature	Dental and orthopedic lattice implants
Inconel 718 (Nickel Superalloy)	Tensile strength: ~1000–1300 MPa; Excellent creep resistance (>700°C)	High-temperature stability; oxidation resistant	High-temperature structural metamaterials, turbine lattice components
Inconel 625	Tensile strength: ~800–1000 MPa; Excellent corrosion resistance	Stable in aggressive environments	Marine and aerospace metamaterial structures
CP-Titanium (Commercially Pure Ti)	Moderate strength; Excellent corrosion resistance; Highly biocompatible	Excellent compatibility in vacuum; Good ductility	Biomedical porous scaffolds, lightweight lattice implants
Ti-5553 Alloy	High strength; Good fracture toughness	Superior load-bearing capability	High-performance aerospace metamaterials
Nickel-Based Superalloys (Various Grades)	High thermal resistance; Good mechanical stability at elevated temperature	Suitable for high-temperature vacuum processing	Aerospace and energy-sector lattice metamaterials

Material selection in EBM-based mechanical metamaterials depends on the strength requirements, density, thermal stability, fatigue resistance, and application environment. The elevated build temperature and vacuum environment of EBM make it particularly suitable for producing structurally reliable metallic lattice metamaterials for aerospace and biomedical engineering applications.

Table 2.8 compares the three major metal powder bed fusion technologies used in mechanical metamaterial fabrication. SLM and DMLS are similar in their operational principle, both using high-power lasers in inert gas environments to produce dense metallic parts with high mechanical strength. EBM differs primarily in its use of an electron beam and a vacuum chamber, along with elevated build temperatures that significantly reduce residual stresses.

TABLE 2.8
Comparative Analysis of SLM, DMLS, and EBM Processes

Parameter	Selective Laser Melting (SLM)	Direct Metal Laser Sintering (DMLS)	Electron Beam Melting (EBM)
Energy Source	High-power fiber laser	Fiber laser	Electron beam
Processing Environment	Inert gas atmosphere (Argon/Nitrogen)	Inert gas atmosphere	High-vacuum environment
Powder Fusion Mechanism	Complete melting of metal powder	Near-complete melting (historically termed sintering)	Complete melting with preheating
Build Temperature	Moderate	Moderate	High (preheated powder bed)
Residual Stress	Relatively high (requires stress relief)	Moderate to high	Lower due to elevated build temperature
Material Compatibility	Wide range of metals (Ti, Al, Steel, Inconel, etc.)	Similar to SLM (steel, Ti, Al, superalloys)	Primarily Ti alloys, Co-Cr, Ni superalloys
Surface Finish	Moderate (powder-based roughness)	Similar to SLM	Slightly rougher due to larger powder size
Dimensional Accuracy	High	High	Moderate–High
Mechanical Properties	Near fully dense; high strength	Near fully dense; comparable to SLM	Dense parts; good ductility and low internal stress
Support Structures	Required for overhangs	Required	Required (but easier removal due to preheat sintering)
Post-Processing	Heat treatment, machining, support removal	Heat treatment, machining	Heat treatment (often reduced stress relief needed)
Equipment Cost	High	High	Very high (vacuum system required)
Typical Applications	Aerospace lattices, lightweight structural metamaterials	Biomedical implants, tooling, structural components	Titanium implants, aerospace load-bearing lattices

SLM and DMLS offer broader material compatibility and high dimensional precision, making them suitable for diverse structural applications. EBM, while more specialized and expensive, is particularly advantageous for titanium-based metamaterials due to its reduced thermal distortion and improved structural reliability.

2.4 FABRICATION TECHNIQUES

2.4.1 Microfabrication and Lithography

Microfabrication and lithography-based techniques are advanced manufacturing approaches used to fabricate micro- and nanoscale mechanical metamaterials with extremely fine geometric features. These techniques, originally developed for semiconductor and microelectromechanical systems (MEMS) fabrication, enable precise control over feature size, structural thickness, and pattern fidelity. In mechanical metamaterials, such a high-resolution capability is essential for realizing micro-lattices, phononic crystals, and architected materials designed to operate at small length scales.

2.4.1.1 Working Principle

Microfabrication involves a sequence of material deposition, patterning, etching, and removal processes to create microstructured components. Lithography, a key step in this process, uses light or electron beams to transfer geometric patterns from a mask onto a photosensitive material (photoresist). The patterned photoresist then serves as a template for selective material removal or deposition, enabling the creation of intricate structural features.

2.4.1.2 Common Microfabrication Techniques

1. **Photolithography**: A light-based pattern transfer process widely used for planar microstructures. It is suitable for producing two-dimensional periodic architectures and layered metamaterials.
2. **Soft Lithography**: Utilizes elastomeric molds (commonly PDMS) to replicate microscale patterns. It is particularly useful for fabricating soft mechanical metamaterials and biomedical scaffolds.
3. **Electron Beam Lithography (EBL)**: Employs a focused electron beam for the direct writing of nanoscale features with extremely high resolution, often below 100 nm.
4. **Two-Photon Polymerization (2PP)**: A laser-based microfabrication technique that enables true three-dimensional microstructures with sub-micron precision. This method is widely used for micro-lattice metamaterials and hierarchical architectures.
5. **MEMS-Based Fabrication**: Integrates deposition, etching, and micromachining processes to produce microscale mechanical structures with precise mechanical functionality.

Advantages

- Extremely high resolution and geometric precision
- Capability to fabricate micro- and nanoscale architectures
- Suitable for high-frequency acoustic and phononic metamaterials
- Excellent dimensional control

Limitations

- Limited scalability for large structures
- Higher fabrication cost
- Complex and multi-step processing
- Material selection constraints

2.4.1.3 Photolithography

Photolithography is a microfabrication technique used to transfer geometric patterns onto a substrate through the controlled exposure of a photosensitive material (photoresist) to ultraviolet (UV) light. Originally developed for semiconductor manufacturing, photolithography has become an essential method for fabricating microscale mechanical metamaterials, particularly planar and layered architectures.

2.4.1.3.1 Working Principle

The process begins with the preparation of a clean substrate, typically silicon, glass, or metal. A thin layer of photoresist is deposited onto the substrate through spin coating, forming a uniform coating. A photomask, containing the desired geometric pattern, is aligned over the coated substrate. When exposed to UV light, the photoresist undergoes a chemical transformation in the illuminated regions.

Classification according to the photoresist type:

- **Positive photoresist**: The exposed regions become soluble and are removed during development.
- **Negative photoresist**: The exposed regions polymerize and remain after development.

After development, the patterned photoresist acts as a protective mask for subsequent processes such as etching, deposition, or electroplating. Once the structural pattern is transferred into the substrate material, the remaining photoresist is removed.

Process Steps

1. Substrate cleaning
2. Photoresist spin coating
3. Soft baking (to remove solvents)
4. Mask alignment and UV exposure
5. Development of photoresist
6. Etching or material deposition
7. Photoresist stripping

2.4.1.3.2 Resolution and Capabilities

Photolithography can achieve feature sizes down to the micron scale, depending on the wavelength and system capability. Advanced deep-UV systems can produce even smaller features. However, traditional photolithography is primarily suited for **two-dimensional or layered structures**, making it particularly useful for planar metamaterials.

Advantages

- High precision and repeatability
- Suitable for batch fabrication
- Well-established and scalable process
- Compatible with MEMS integration

Limitations

- Primarily limited to planar geometries
- Requires cleanroom facilities
- Mask fabrication can be expensive
- Limited to materials compatible with lithographic processing

Figure 2.15 illustrates the sequential stages involved in the photolithography process used for the microfabrication of mechanical metamaterials. The process begins with a clean substrate, onto which a thin photoresist layer is uniformly deposited. A patterned photomask is then aligned above the coated substrate. When exposed to UV light, the radiation passes through the transparent regions of the mask and chemically modifies the exposed portions of the photoresist.

Following exposure, the substrate undergoes development, during which either the exposed or unexposed regions of the photoresist are removed, depending on whether a positive or negative resist is used. The remaining patterned photoresist acts as a protective mask for the subsequent etching step, where material from the uncovered substrate regions is selectively removed. After etching, the residual photoresist is stripped, resulting in the final patterned structure transferred onto the substrate.

The figure clearly demonstrates the mask-based pattern transfer mechanism that enables high-precision microstructural fabrication. This process is particularly important for producing planar phononic crystals, thin-film metamaterials, and MEMS-integrated mechanical metamaterial architectures.

2.4.1.4 Soft Lithography

Soft lithography is a microfabrication technique that employs elastomeric stamps, molds, or templates to transfer micro- and nanoscale patterns onto substrates. Unlike

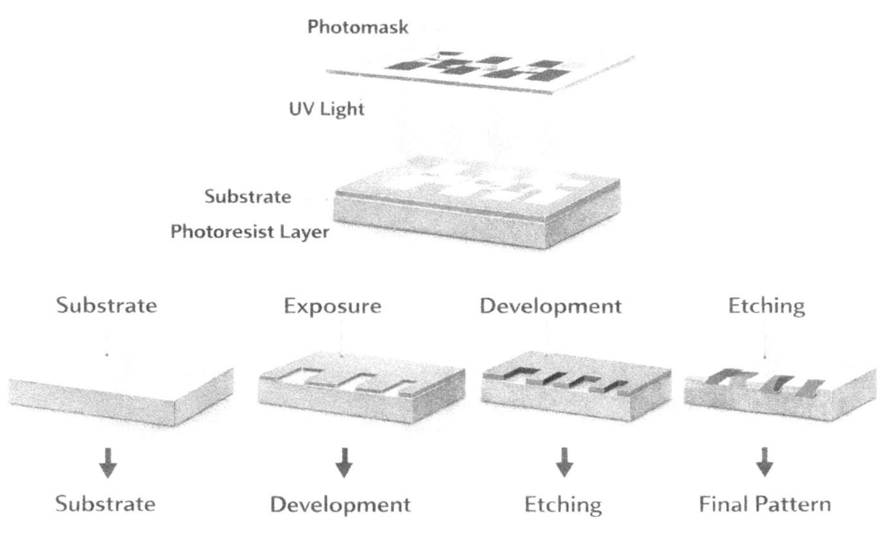

FIGURE 2.15 Schematic of photolithographic patterning steps.

conventional photolithography, which relies on rigid photomasks and complex cleanroom processes, soft lithography uses flexible polymeric materials, most commonly polydimethylsiloxane (PDMS) to replicate and pattern microstructures. This technique is widely used for fabricating soft mechanical metamaterials, biomedical scaffolds, and flexible micro-architected systems.

2.4.1.4.1 Working Principle

Soft lithography is typically a replication-based process. It begins with the fabrication of a master mold, often created using photolithography. A liquid elastomer (e.g., PDMS) is poured over the master and cured to form a flexible stamp containing the negative of the desired pattern. This elastomeric stamp is then used to transfer or imprint the pattern onto another substrate through methods such as molding, microcontact printing, or embossing.

Because PDMS is flexible and conformable, it can replicate fine surface details with high fidelity and can conform to slightly curved or uneven surfaces.

2.4.1.4.2 Common Soft Lithography Techniques

1. **Microcontact Printing (μCP)**: Uses a PDMS stamp to transfer patterned molecules or inks onto a substrate.
2. **Replica Molding**: Replicates microstructures by casting polymer against a patterned mold.
3. **Microtransfer Molding (μTM)**: Transfers patterned polymer structures onto a substrate.
4. **Embossing**: Presses patterned molds into thermoplastic materials under heat and pressure.

Materials Used

- Polydimethylsiloxane (PDMS)
- Polyurethane
- Hydrogels
- Silicone-based elastomers
- Thermoplastic polymers

These materials are particularly suitable for soft and flexible mechanical metamaterials.

Advantages

- Cost-effective compared to traditional microfabrication
- Suitable for soft and flexible materials
- Capable of replicating microscale features
- Does not always require complex cleanroom facilities

Limitations

- Limited structural rigidity
- Reduced precision compared to advanced lithographic methods

- Primarily suited for surface and planar structures
- Mold degradation over repeated use

2.4.1.5 Electron Beam Lithography (EBL)

EBL is a high-resolution micro- and nanofabrication technique that uses a focused beam of electrons to directly write patterns onto an electron-sensitive resist. Unlike photolithography, which relies on optical masks and UV light, EBL is a maskless, direct-write process capable of achieving feature sizes in the nanometer range. This makes it particularly suitable for fabricating nanoscale mechanical and phononic metamaterials, where geometric precision at extremely small scales is essential.

2.4.1.5.1 Working Principle

In EBL, a substrate is first coated with an **electron-sensitive resist**, such as polymethyl methacrylate (PMMA). A finely focused electron beam, generated by an electron gun and controlled through electromagnetic lenses and scanning systems, is directed across the resist surface according to a digitally defined pattern.

The interaction of the high-energy electrons with the resist causes localized chemical modifications:

- **Positive resist**: The exposed regions become soluble and are removed during development.
- **Negative resist**: The exposed regions cross-link and remain after development.

Following development, the patterned resist serves as a mask for subsequent processes such as etching, metal deposition, or lift-off. The final structure is then transferred onto the underlying substrate.

2.4.1.5.2 Resolution and Capabilities

EBL is capable of producing features as small as 10–20 nanometers, depending on system parameters and resist type. Because it does not require a photomask, EBL offers exceptional flexibility in pattern design and the rapid modification of geometries. However, the process is relatively slow in comparison with optical lithography due to its serial writing mechanism.

Materials Used

- Electron-sensitive resists (PMMA, HSQ)
- Silicon substrates
- Metal thin films
- Dielectric materials

Advantages

- Extremely high resolution (nanoscale fabrication)
- Maskless patterning (flexible design modification)
- Suitable for research and prototyping
- High geometric precision

Limitations

- Slow writing speed (serial process)
- High equipment cost
- Limited scalability for large-area fabrication
- Requires vacuum and cleanroom environment

2.4.1.6 Two-Photon Polymerization (2PP)

Two-Photon Polymerization (2PP) is an advanced laser-based micro- and nanofabrication technique that enables the direct writing of complex three-dimensional microstructures with sub-micron resolution. Unlike conventional photolithography, which is largely limited to planar geometries, 2PP allows true volumetric fabrication within a photosensitive resin. This capability makes it highly suitable for producing micro-lattice mechanical metamaterials, hierarchical architectures, and nanoscale unit cells.

2.4.1.6.1 Working Principle

The 2PP process is based on a nonlinear optical phenomenon known as two-photon absorption. A femtosecond-pulsed laser is tightly focused into a photosensitive resin. Polymerization occurs only at the focal point where the photon density is sufficiently high for two photons to be absorbed simultaneously by a photoinitiator molecule.

Because two-photon absorption requires a very high intensity, polymerization is confined to a tiny, three-dimensional voxel (volumetric pixel) around the laser focus. By scanning the focal point in three dimensions using precision motion stages, complex microstructures can be written directly inside the resin.

After writing is complete, the structure is developed by removing the uncured resin, leaving behind a fully formed 3D microstructure.

Process Steps

1. Preparation of photosensitive resin
2. Laser focusing and voxel formation
3. Controlled 3D scanning of focal point
4. Selective polymerization within the resin
5. Development and removal of unpolymerized material

2.4.1.6.2 Resolution and Capabilities

2PP can achieve feature sizes below 100 nanometers, depending on system parameters and resin chemistry. The technique supports the following:

- True 3D free-form geometries
- Complex micro-lattices
- Hierarchical and graded structures
- Nanoscale mechanical resonators

Materials Used

- Photosensitive polymer resins
- Hybrid organic–inorganic resins
- Functionalized resins (conductive, biocompatible, etc.)

Advantages

- Ultra-high resolution (sub-micron to nanoscale)
- True 3D fabrication capability
- Excellent geometric precision
- Suitable for complex hierarchical metamaterials

Limitations

- Limited build volume
- Slow fabrication speed
- High equipment cost
- Material restrictions (primarily photopolymers)

2.4.1.7 MEMS-Based Fabrication

MEMS-based fabrication refers to microfabrication techniques originally developed for microelectromechanical systems (MEMS) that are adapted to create microscale mechanical metamaterials. These techniques integrate material deposition, lithographic patterning, etching, and micromachining processes to produce highly precise microstructures with controlled mechanical functionality. MEMS fabrication is particularly relevant for metamaterials operating at microscale dimensions, especially in applications involving micro-actuation, sensing, and high-frequency mechanical response.

2.4.1.7.1 Working Principle

MEMS fabrication relies on sequential microfabrication steps performed on substrates such as silicon wafers. The process typically involves the following:

1. **Thin-Film Deposition**: Materials such as polysilicon, silicon nitride, or metals are deposited using techniques like chemical vapor deposition (CVD) or physical vapor deposition (PVD).
2. **Lithographic Patterning**: Photolithography defines the geometric pattern on the deposited layer.
3. **Etching**: Wet or dry etching selectively removes material to form structural features.
4. **Release Process**: Sacrificial layers are removed to free the movable microstructures.

By combining multiple layers and etching steps, three-dimensional micro-architectures can be fabricated with high dimensional accuracy.

Common MEMS Fabrication Techniques

- **Bulk Micromachining**: Removes material from the substrate to create cavities or structural elements.
- **Surface Micromachining**: Builds structures layer-by-layer on the surface using sacrificial layers.
- **Deep Reactive Ion Etching (DRIE)**: Enables high-aspect-ratio microstructures in silicon.
- **Wafer Bonding**: Joins multiple patterned wafers to form complex 3D structures.

Materials Used

- Silicon (single-crystal, polysilicon)
- Silicon dioxide (SiO_2)
- Silicon nitride (Si_3N_4)
- Thin metal films (Al, Au, Pt)
- Piezoelectric materials (PZT, AlN)

Advantages

- Extremely high dimensional precision
- Excellent repeatability and batch processing capability
- Integration of mechanical and electronic functions
- Suitable for microscale dynamic metamaterials

Limitations

- Primarily limited to small-scale fabrication
- Complex, multi-step cleanroom processes
- Material restrictions (mainly silicon-based systems)
- High infrastructure cost

2.4.2 Laser Cutting and Precision Machining

Laser cutting and precision machining are subtractive manufacturing techniques widely used for fabricating planar and sheet-based mechanical metamaterials. These processes are particularly effective for producing architected structures derived from thin plates, sheets, or layered assemblies, including auxetic patterns, kirigami-inspired designs, and compliant mechanisms.

2.4.2.1 Laser Cutting

Laser cutting employs a focused high-energy laser beam to cut material along pre-defined paths derived from CAD data. The laser locally melts, burns, or vaporizes material, producing precise cuts with minimal mechanical contact. The process is computer-controlled (computer numerical control (CNC)-based), allowing high repeatability and geometric accuracy.

Laser cutting is especially suitable for:

- 2D periodic and aperiodic metamaterial patterns
- Auxetic re-entrant structures
- Kirigami and origami-inspired architectures
- Thin metallic or polymer sheet metamaterials.

Common materials processed include stainless steel, aluminum, acrylic, polymers, and composite sheets.

Advantages

- High cutting precision
- Rapid production for sheet geometries
- No mechanical tool wear
- Suitable for complex planar geometries

Limitations

- Primarily restricted to 2D or layered structures
- Limited capability for complex 3D internal lattices
- Heat-affected zone (HAZ) may influence local mechanical properties

2.4.2.1.1 Precision Machining (CNC Machining and Micromachining)

Precision machining involves the controlled removal of material using cutting tools such as milling cutters, drills, and lathes. CNC systems enable high-accuracy fabrication of complex geometries in metals, polymers, and composites.

Micromachining techniques extend these capabilities to microscale feature fabrication, enabling the creation of fine structural details required in mechanical metamaterials.

Precision machining is suitable for:

- 3D lattice components (limited complexity)
- High-strength metallic metamaterials
- Prototype validation of topology-optimized structures
- Structural metamaterials requiring tight tolerances

Advantages

- High dimensional accuracy
- Suitable for high-strength materials
- Excellent surface finish (with finishing operations)
- Reliable for load-bearing components

Limitations

- Material wastage (subtractive process)
- Limited capability for internal closed-cell architectures
- Tool wear and machining constraints

2.4.3 Hybrid and Multimaterial Fabrication

Hybrid and multimaterial fabrication refers to advanced manufacturing strategies that integrate multiple fabrication processes and/or multiple materials within a single structure to achieve enhanced mechanical performance and multifunctionality. In the context of mechanical metamaterials, such approaches enable the spatial variation of stiffness, strength, density, damping, or actuation capability within a unified architected system.

2.4.3.1 Hybrid Fabrication

Hybrid fabrication combines two or more manufacturing techniques in a coordinated workflow. For example:

- AM followed by CNC machining for improved surface finish and dimensional accuracy.
- Metal AM integrated with post-processing heat treatment.
- 3D-printed lattice cores bonded with conventionally manufactured face sheets (sandwich structures).

Hybrid processing is particularly beneficial when the design demands geometric complexity along with high structural precision or surface quality. In mechanical metamaterials, hybrid techniques allow the fabrication of optimized internal lattices while ensuring that load-bearing outer surfaces meet engineering standards.

2.4.3.2 Multimaterial Fabrication

Multimaterial fabrication involves the use of two or more materials within a single component. This can be achieved through:

- Multi-nozzle extrusion in polymer AM.
- Material jetting systems capable of depositing materials with varying stiffness.
- Functionally graded material (FGM) deposition.
- Embedded reinforcement fibers or smart materials within a lattice framework.

This approach enables the creation of metamaterials with spatially programmable mechanical behavior, such as graded stiffness, selective compliance, or integrated sensing capability.

2.4.3.3 Design Advantages

Hybrid and multimaterial strategies enable:

- Functionally graded architectures
- Improved energy absorption and impact resistance
- Tunable mechanical response
- Integration of active elements (e.g., piezoelectric or shape memory components)
- Enhanced durability and performance optimization.

These approaches are particularly useful in aerospace, biomedical implants, soft robotics, and adaptive architectural systems.

2.4.3.4 Challenges

Despite their advantages, hybrid and multimaterial fabrication present challenges including:

- Interfacial bonding between dissimilar materials
- Differential thermal expansion
- Process compatibility and sequencing
- Increased design and manufacturing complexity
- Higher production costs.

Careful material selection and computational design integration are essential to ensure mechanical reliability and long-term performance.

2.5 TOPOLOGICAL OPTIMIZATION

Topology optimization is a computational design methodology used to determine the optimal material distribution within a defined design domain to achieve specified performance objectives under given constraints. In mechanical metamaterials, topology optimization plays a critical role in generating architected unit cells and lattice geometries that exhibit tailored mechanical properties such as a high stiffness-to-weight ratio, a negative Poisson's ratio, energy absorption capability, or controlled deformation modes.

2.5.1 Fundamental Concept

Unlike traditional shape optimization, which modifies predefined geometric boundaries, topology optimization allows material to be added or removed freely within the design space. The process begins with the following:

- A prescribed design domain
- Defined loading and boundary conditions
- Objective functions (e.g., maximize stiffness, minimize compliance)
- Constraints (e.g., volume fraction, stress limits)

Through iterative numerical algorithms, the method determines the optimal material layout that satisfies these requirements.

2.5.2 Mathematical Framework

Most topology optimization problems are formulated using the below:

- **Compliance minimization** (maximize stiffness for a given material volume)
- **Volume constraint optimization**
- **Stress-based optimization**
- **Multi-objective optimization**

The Solid Isotropic Material with Penalization (SIMP) method is one of the most commonly used approaches. In SIMP, each finite element is assigned a density variable between 0 and 1, representing void or solid material, and penalization functions drive the solution toward a clear solid–void distribution.

2.6 BIO-INSPIRED AND FRACTAL DESIGNS

Bio-inspired and fractal design strategies draw inspiration from natural hierarchical systems and self-similar geometric patterns to enhance the mechanical performance of metamaterials. These approaches leverage principles observed in biological materials—such as bone, nacre, wood, and spider silk—and mathematical fractal geometries to achieve superior strength-to-weight ratios, energy absorption, toughness, and multifunctionality.

2.6.1 Bio-Inspired Design

Bio-inspired mechanical metamaterials emulate structural features found in nature, where mechanical efficiency is achieved through hierarchical organization, gradient distribution, and optimized load paths. Natural materials often combine stiffness and toughness through multiscale architecture rather than through material composition alone.

Key bio-inspired concepts include the following:

- **Hierarchical structuring**: Multiple length scales contributing to overall mechanical performance.
- **Cellular architectures**: Honeycomb and trabecular bone-like lattices.
- **Layered composites**: Nacre-inspired brick-and-mortar configurations.
- **Gradient structures**: Functionally graded properties similar to biological tissues.

In mechanical metamaterials, bio-inspired architectures are used to design lightweight structures with enhanced impact resistance, fracture toughness, and fatigue life.

2.6.2 Fractal Design

Fractal designs are based on self-similar geometric patterns that repeat across multiple scales. Fractals exhibit scale invariance, meaning that similar structural motifs appear at different magnifications. Incorporating fractal geometries into mechanical metamaterials enables the following:

- Improved stiffness-to-weight efficiency
- Enhanced energy dissipation
- Increased surface area-to-volume ratio
- Tunable mechanical and dynamic response

Fractal-based metamaterials often use recursive lattice patterns, Koch-like geometries, or branching structures that distribute stresses efficiently and improve mechanical robustness.

Advantages of Bio-Inspired and Fractal Approaches

- Efficient load distribution
- High mechanical strength with reduced material usage
- Improved damage tolerance

- Multiscale energy absorption
- Enhanced structural stability

2.7 SMART MATERIALS AND PROGRAMMABLE METAMATERIALS

Smart materials and programmable metamaterials represent an advanced class of architected systems capable of adapting their mechanical response in response to external stimuli. Unlike passive mechanical metamaterials whose properties are fixed by geometry, these systems integrate responsive materials or embedded actuation mechanisms to enable tunable, reconfigurable, or time-dependent behavior.

2.7.1 Smart Materials

Smart materials are materials that exhibit a measurable and reversible change in properties when exposed to external stimuli such as temperature, an electric field, a magnetic field, stress, or moisture. In mechanical metamaterials, smart materials are often incorporated within architected structures to enable controlled deformation or stiffness modulation.

Common smart materials include:

- **Shape Memory Alloys (SMAs)**: Exhibit shape recovery upon heating due to phase transformation.
- **Shape Memory Polymers (SMPs)**: Recover programmed shapes when subjected to thermal activation.
- **Piezoelectric Materials**: Generate mechanical strain under an applied electric field.
- **Magnetorheological (MR) and Electrorheological (ER) Materials**: Change stiffness or viscosity under magnetic or electric fields.
- **Hydrogels**: Swell or contract in response to moisture or pH changes.

These materials enable mechanical metamaterials to transition between multiple stable states, adjust stiffness, or respond dynamically to environmental conditions.

2.7.2 Programmable Metamaterials

Programmable metamaterials are architected systems whose mechanical behavior can be predefined, tuned, or reconfigured through structural design and/or external control. Programmability may arise from the following:

- Geometric reconfiguration (origami and kirigami architectures)
- Multistable unit cells
- Embedded actuation elements
- Spatially graded material distribution
- Controlled instability mechanisms

These structures can switch between configurations with distinct stiffness, shape, or load-bearing characteristics, enabling property modulation without altering the base material.

2.7.3 Mechanisms of Programmability

Programmable behavior in mechanical metamaterials is typically achieved through

- Elastic instability (snap-through buckling)
- Phase transformation (in SMAs)
- Differential thermal expansion
- Magnetic or electric actuation
- Mechanical locking/unlocking mechanisms

Such mechanisms allow reversible transitions between stable configurations, enabling adaptive performance.

Applications

Smart and programmable metamaterials are widely explored in:

- Soft robotics and adaptive grippers
- Deployable aerospace structures
- Vibration isolation systems with tunable frequency response
- Biomedical devices and stents
- Adaptive architectural façades.

Challenges

- Integration of active materials without compromising structural integrity
- Energy requirements for actuation
- Long-term durability and fatigue resistance
- Increased design and fabrication complexity

2.8 SUMMARY

This chapter comprehensively addressed the design principles and fabrication techniques of mechanical metamaterials, highlighting how their mechanical behavior is governed primarily by engineered architecture rather than intrinsic material properties. It began with core design concepts such as unit cell–based design, geometry and topology control, and architectural classifications including periodic, aperiodic, and graded structures, which enable precise tuning of stiffness, deformation modes, and energy absorption. Advanced strategies such as multistability, programmability, hierarchical and multiscale designs, and computational topology optimization were discussed as powerful tools for achieving adaptive, high-efficiency, and non-intuitive structural performance. This chapter further examined fabrication routes essential for translating designs into physical structures, including polymer-based AM (FDM, SLA, SLS), metal AM (SLM, DMLS, EBM), and microscale approaches such as photolithography, soft lithography, EBL, two-photon polymerization, and MEMS-based fabrication, along with laser cutting, precision machining, and hybrid multimaterial strategies. Finally, emerging directions such as bio-inspired and fractal architectures and the integration of smart materials for programmable metamaterials were presented, emphasizing the convergence of computational design, advanced manufacturing, and material innovation in developing next-generation architected mechanical systems.

REVIEW QUESTIONS

1 Define mechanical metamaterials and explain how their properties differ from conventional materials.
2 What is unit cell–based design in mechanical metamaterials?
3 How does geometry influence the mechanical behavior of metamaterials?
4 Explain the role of topology in architected materials.
5 Distinguish between stretching-dominated and bending-dominated architectures.
6 What are periodic architectures, and what are their advantages?
7 Explain the significance of aperiodic architectures.
8 How do graded architectures enhance mechanical performance?
9 Define multistability in mechanical metamaterials.
10 What mechanisms enable programmability in metamaterials?
11 Explain hierarchical and multiscale design concepts.
12 How does scale influence effective material properties?
13 What is compliance minimization in topology optimization?
14 Explain the SIMP method used in topology optimization.
15 Discuss the importance of computational modeling in metamaterial design.
16 What are the challenges in topology optimization for manufacturable designs?
17 How do bio-inspired designs improve structural efficiency?
18 Explain fractal-based mechanical metamaterials.
19 What are the benefits of incorporating smart materials into metamaterials?
20 Describe the role of instability (buckling/snap-through) in programmable metamaterials.
21 Explain the working principle of Fused Deposition Modeling (FDM).
22 What are the limitations of FDM for mechanical metamaterials?
23 Describe the working principle of SLA.
24 Compare SLA and FDM in terms of resolution and mechanical properties.
25 Explain the SLS process.
26 What advantages does SLS offer over FDM?
27 Discuss the importance of support structures in additive manufacturing.
28 What are the main differences between SLM and SLS?
29 Explain the principle of SLM.
30 Describe the DMLS process.
31 Compare SLM and DMLS processes.
32 Explain the working principle of EBM.
33 Why is EBM particularly suitable for titanium alloys?
34 Discuss residual stresses in metal additive manufacturing.
35 What post-processing steps are required in metal AM?
36 What is photolithography and how is it used in metamaterial fabrication?
37 Explain the difference between positive and negative photoresists.
38 Describe the principle of soft lithography.

39 What is Electron Beam Lithography (EBL), and what is its advantage?
40 Explain the working mechanism of Two-Photon Polymerization (2PP).
41 What are the advantages of 2PP for micro-lattice fabrication?
42 Describe MEMS-based fabrication methods.
43 What are the limitations of microfabrication techniques?
44 Explain the role of laser cutting in planar metamaterial fabrication.
45 How does precision machining contribute to metamaterial development?
46 What is hybrid fabrication, and why is it important?
47 Explain multimaterial fabrication and its advantages.
48 Discuss functionally graded mechanical metamaterials.
49 How does computational optimization integrate with additive manufacturing?
50 Discuss future trends in smart and programmable mechanical metamaterials.

3 Mechanical Properties and Functional Behavior

3.1 INTRODUCTION

Mechanical metamaterials exhibit unique mechanical properties and functional behavior that arise primarily from their engineered internal architecture rather than their base material composition. By carefully designing unit-cell geometry and structural topology, these materials can display unconventional responses, such as negative Poisson's ratio, controlled elastic instabilities, negative stiffness, tunable stiffness, and enhanced energy absorption. Their mechanical performance is governed by mechanisms including bending, rotation, buckling, and snap-through transitions, enabling adaptive, programmable, and lightweight structural solutions. Understanding the relationship between geometry and deformation behavior is essential for predicting effective properties and optimizing performance under static and dynamic loading conditions. This chapter, therefore, establishes the fundamental connection between architectural design and emergent mechanical functionality in advanced metamaterial systems.

3.2 AUXETIC MATERIALS AND NEGATIVE POISSON'S RATIO

Auxetic materials represent a unique class of mechanical metamaterials characterized by a negative Poisson's ratio, meaning they expand laterally when stretched and contract laterally when compressed. This behavior contrasts with conventional materials, which typically become thinner when stretched. The unusual deformation response of auxetic materials arises primarily from their engineered internal geometry, rather than the intrinsic properties of the base material. By designing specific unit-cell architectures such as re-entrant honeycombs, rotating units, chiral structures, and perforated sheets, the structural elements rotate, hinge, or unfold during deformation, leading to lateral expansion under tensile loading (Lim, 2015).

Poisson's ratio (ν) is defined as the negative ratio of transverse strain to axial strain. For most conventional materials, Poisson's ratio ranges between 0 and 0.5, whereas auxetic materials exhibit values less than zero. This negative Poisson's ratio results in several enhanced mechanical properties, including improved shear resistance, higher indentation resistance, enhanced energy absorption, and greater fracture toughness. Additionally, auxetic structures demonstrate synclastic curvature, allowing them to form dome-like surfaces when bent.

Auxetic metamaterials can be realized using various geometric mechanisms. Re-entrant structures utilize inwardly angled cell ribs that rotate outward under tension, causing expansion. Rotating rigid unit mechanisms involve interconnected rigid shapes that rotate relative to each other during deformation. Chiral and anti-chiral

DOI: 10.1201/9781003783671-3

architectures rely on node rotation and ligament bending, while perforated or missing-rib designs introduce strategically placed cuts that allow structural elements to hinge and expand. These deformation mechanisms enable the design of materials with tunable and programmable mechanical behavior.

Due to their unique properties, auxetic materials have attracted significant interest in a wide range of engineering applications. In biomedical engineering, they are used for implants and orthopedic supports due to their ability to conform to complex shapes and distribute loads evenly. In protective equipment, auxetic structures improve impact resistance and energy absorption. Aerospace structures, flexible electronics, and smart textiles also benefit from the enhanced mechanical performance and adaptability of auxetic metamaterials. Consequently, auxetic materials represent a key example of how geometric design can be used to achieve unconventional mechanical behavior in advanced engineering systems.

3.2.1 Fundamentals of Poisson's Ratio

Poisson's ratio is a fundamental mechanical property that describes the relationship between axial deformation and transverse deformation in a material subjected to mechanical loading. When a material is stretched in one direction, it generally experiences a contraction in the perpendicular direction, and when compressed, it expands laterally. This phenomenon is quantified by Poisson's ratio (ν), which is defined as the negative ratio of transverse strain to axial strain:

$$\nu = -\frac{\varepsilon_{transverse}}{\varepsilon_{axial}}$$

where:

ε_{axial} represents the strain along the direction of applied load, and
$\varepsilon_{transverse}$ represents the strain perpendicular to the applied load.

For most conventional engineering materials, Poisson's ratio falls within a specific range. Metals typically exhibit values between 0.25 and 0.35, polymers range between 0.30 and 0.45, and elastomers such as rubber approach 0.5, indicating nearly incompressible behavior. Materials with $\nu = 0$ do not exhibit lateral deformation during axial loading, while $\nu = 0.5$ represents an ideal incompressible material where volume remains constant during deformation (Norris, 2006).

3.2.2 Mechanisms of Auxetic Behavior

Auxetic behavior in mechanical metamaterials arises primarily from engineered structural geometry and deformation mechanisms rather than intrinsic properties of the base material. Unlike conventional materials, where deformation is dominated by atomic bond stretching, auxetic materials rely on rotational, hinging, and bending mechanisms within their internal architecture. These mechanisms enable lateral expansion when the material is stretched and lateral contraction when compressed, resulting in a negative Poisson's ratio.

One of the most widely studied mechanisms is the re-entrant geometry, in which the cell ribs are inclined inward toward the cell center. When tensile load is applied, these inwardly angled ribs rotate outward, causing the structure to expand laterally. Another common mechanism is the rotating unit mechanism, where rigid elements such as squares, triangles, or polygons are connected at flexible hinges. Under tension, these units rotate relative to each other, leading to an increase in both axial and transverse dimensions.

Chiral and anti-chiral structures represent another class of auxetic designs. These architectures consist of circular nodes connected by ligaments. When load is applied, the nodes rotate and the ligaments bend, producing lateral expansion. Similarly, missing-rib and perforated sheet structures achieve auxetic behavior through strategically placed cuts or voids that allow structural elements to rotate or hinge during deformation.

In all these mechanisms, the negative Poisson's ratio is achieved through geometric rearrangement of structural elements, including rotation, bending, and hinging, rather than through direct stretching of the material itself. By carefully controlling the geometry, connectivity, and stiffness of these elements, engineers can design metamaterials with tunable auxetic responses and enhanced mechanical performance. These mechanisms form the fundamental basis for many advanced auxetic metamaterial designs used in aerospace, biomedical, and protective engineering applications.

Figure 3.1 illustrates the principal geometric mechanisms responsible for auxetic behavior in mechanical metamaterials, where structures exhibit a negative Poisson's ratio. The figure presents four common structural designs: re-entrant geometry, rotating

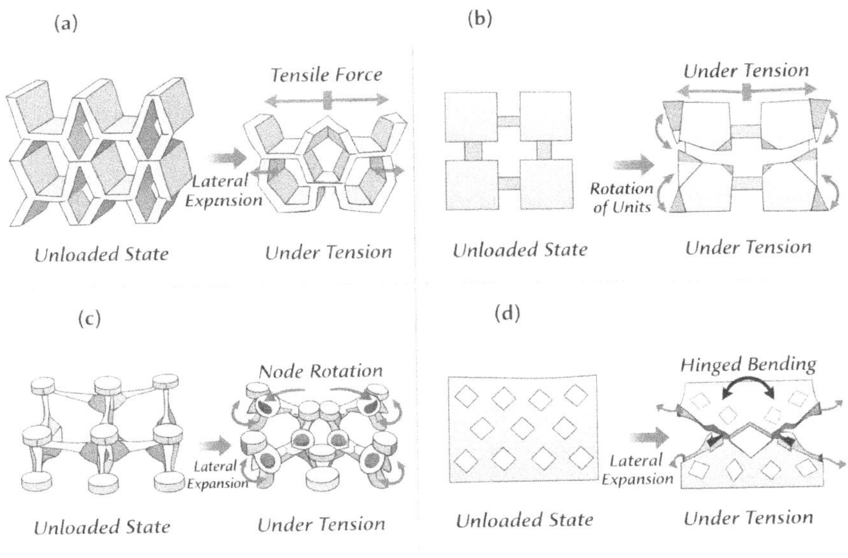

FIGURE 3.1 Mechanisms of auxetic behavior in mechanical metamaterials. (a) Re-entrant Geometry (b) Rotating Unit Mechanism (c) Chiral Structure (d) Perforated Sheet Structure.

unit mechanism, chiral structure, and perforated sheet structure, each shown in both the unloaded state and under tensile loading. In the re-entrant geometry, inwardly inclined cell ribs rotate outward when tensile force is applied, causing lateral expansion of the structure. The rotating unit mechanism consists of rigid units connected at hinges; when tension is applied, the units rotate relative to each other, leading to expansion in both axial and transverse directions. The chiral structure contains circular nodes connected by ligaments, where tensile loading induces node rotation and ligament bending, resulting in lateral expansion. In the perforated sheet structure, strategically placed cuts create hinging segments that bend and rotate during stretching, allowing the sheet to expand laterally. The arrows in the figure indicate the direction of applied forces and deformation mechanisms. Overall, the figure demonstrates how auxetic behavior arises from geometric rearrangement, rotation, and bending of structural elements, rather than from the intrinsic material properties.

3.2.3 Mechanical Advantages of Auxetics

Auxetic materials exhibit several enhanced mechanical properties compared to conventional materials due to their negative Poisson's ratio and unique deformation mechanisms. The lateral expansion that occurs during tensile loading leads to improved structural performance under various mechanical conditions. These advantages arise from the rotation, hinging, and bending of structural elements, which redistribute stresses more efficiently and enhance the overall stability of the structure.

One of the most significant advantages of auxetic materials is their enhanced indentation resistance. When subjected to localized loading or impact, auxetic structures expand laterally around the contact region, increasing local density and stiffness. This behavior improves the material's ability to resist penetration and distribute loads more uniformly. Auxetic materials also exhibit higher shear modulus and improved shear resistance, making them more stable under complex loading conditions.

Another important benefit is their superior energy absorption capability. Due to the deformation mechanisms within their architecture, auxetic materials can dissipate mechanical energy more effectively during compression or impact. This property makes them highly suitable for protective systems such as helmets, body armor, and crashworthy automotive components. Additionally, auxetic materials demonstrate improved fracture toughness, as lateral expansion around crack tips helps reduce stress concentration and slows crack propagation.

Auxetic structures also exhibit synclastic curvature, meaning that when bent, they form dome-like surfaces rather than saddle-shaped surfaces typical of conventional materials. This characteristic enables better conformability to curved surfaces and enhances structural adaptability. As a result, auxetic metamaterials are increasingly applied in fields such as biomedical implants, aerospace structures, flexible electronics, and protective equipment, where enhanced mechanical performance and adaptability are essential.

Figure 3.2 illustrates the key mechanical advantages of auxetic materials that arise from their negative Poisson's ratio and unique deformation mechanisms. The figure presents four major performance benefits: enhanced indentation resistance, improved shear resistance, superior energy absorption, and improved fracture toughness. In the

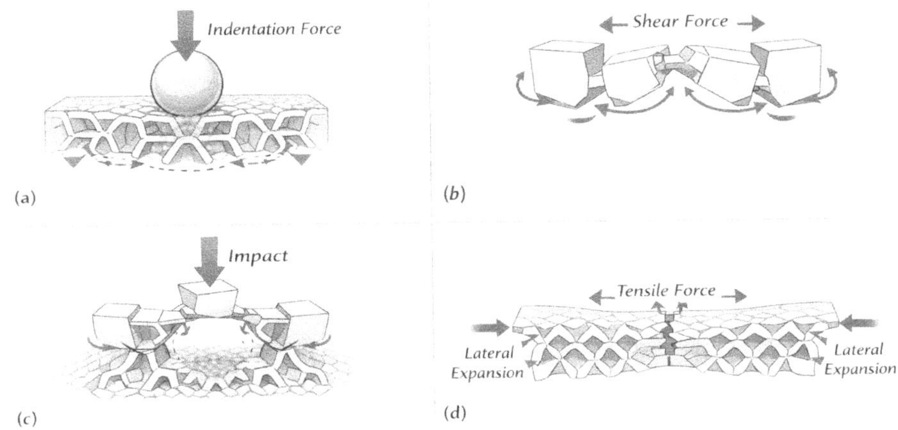

FIGURE 3.2 Mechanical advantages of auxetic materials. (a) Enhanced Indentation Resistance; (b) Improved Shear Resistance; (c) Superior Energy Absorption; (d) Improved Fracture Toughness.

first illustration, the auxetic structure subjected to an indentation force expands laterally around the contact point, increasing local density and stiffness, which enhances resistance to penetration. The second illustration demonstrates improved shear resistance, where the rotation of structural units under shear loading allows the structure to redistribute stresses more effectively, improving structural stability. The third illustration shows superior energy absorption, where the auxetic architecture collapses progressively under impact, dissipating mechanical energy and reducing transmitted forces. Finally, the fourth illustration depicts improved fracture toughness, where lateral expansion near a crack tip reduces stress concentration and slows crack propagation. The arrows in the figure represent applied forces and deformation directions, highlighting how the geometric rearrangement of auxetic unit cells contributes to enhanced mechanical performance.

3.2.4 Analytical and Computational Modeling

Analytical and computational modeling play a crucial role in understanding and predicting the mechanical behavior of auxetic materials and mechanical metamaterials. Since the unique properties of these materials arise primarily from their architectural geometry rather than intrinsic material composition, modeling approaches are essential for evaluating effective mechanical properties, such as Poisson's ratio, elastic modulus, and deformation behavior. Analytical models are often based on simplified representations of the unit-cell geometry, where structural elements are treated as beams or rigid links connected by hinges. Using classical beam theory and kinematic relations, researchers can derive approximate expressions for effective mechanical properties and deformation mechanisms of auxetic structures such as re-entrant honeycombs or rotating unit systems.

However, due to the geometric complexity and nonlinear deformation mechanisms involved in many metamaterial architectures, analytical solutions are often complemented by computational modeling techniques. Finite element analysis (FEA) is widely used to simulate the mechanical response of metamaterial unit cells and periodic structures under various loading conditions. Through numerical simulations, engineers can analyze stress distribution, deformation patterns, buckling behavior, and energy absorption characteristics with high accuracy. Homogenization methods are also employed to determine the effective macroscopic properties of periodic metamaterial structures by averaging the response of representative unit cells.

Computational design tools further enable the exploration of novel architectures through topology optimization and parametric design, allowing engineers to tailor mechanical properties such as stiffness, Poisson's ratio, and energy absorption capacity. These modeling approaches significantly reduce experimental trial-and-error by enabling virtual prototyping and performance evaluation before fabrication. Consequently, analytical and computational modeling form a fundamental foundation for the design, optimization, and application of auxetic mechanical metamaterials in advanced engineering systems.

3.3 ELASTIC INSTABILITIES AND BUCKLING-DRIVEN DESIGNS

Elastic instabilities and buckling-driven designs represent an important class of mechanisms used in mechanical metamaterials to achieve unconventional and programmable mechanical behavior. In conventional structural engineering, buckling is often considered a failure mode that leads to sudden structural collapse. However, in mechanical metamaterials, controlled elastic instabilities are intentionally utilized as functional design features to achieve adaptive deformation, energy absorption, and mechanical switching.

Elastic instability occurs when a structure subjected to compressive or bending loads undergoes a sudden change in deformation configuration after reaching a critical load, commonly referred to as the buckling load. Classical Euler buckling theory describes this phenomenon in slender columns, where instability occurs once the compressive load exceeds a critical value determined by material stiffness, geometric properties, and boundary conditions. In metamaterial architectures, arrays of slender beams, ligaments, or cellular elements are designed such that buckling occurs in a controlled and reversible manner, enabling complex deformation patterns.

Buckling-driven metamaterials often exhibit pattern transformation, where periodic unit-cell geometries undergo significant reconfiguration when subjected to mechanical loading. For example, cellular structures composed of slender beams can transition from a regular configuration to a rotated or collapsed configuration during compression. These transformations can significantly alter the material's effective stiffness, Poisson's ratio, and mechanical anisotropy. In some designs, the post-buckling response leads to bistable or multistable states, where the structure can remain stable in multiple configurations.

Such controlled instability mechanisms enable a wide range of functional behaviors. Buckling-driven metamaterials can exhibit negative stiffness, programmable deformation paths, and tunable mechanical properties. Additionally, snap-through or snap-back instabilities can be exploited for rapid mechanical actuation and energy dissipation. These characteristics are particularly useful in adaptive structures, deployable systems, soft robotics, and vibration control devices.

3.3.1 Fundamentals of Elastic Stability

Elastic stability refers to the ability of a structure to maintain its original equilibrium configuration when subjected to external loads. When the applied load reaches a certain limit, the structure may lose stability and experience a sudden change in deformation pattern, a phenomenon known as buckling. In structural mechanics, elastic stability analysis is essential for understanding the behavior of slender structural elements such as beams, columns, and lattice members. In mechanical metamaterials, these instability phenomena are intentionally controlled and utilized to achieve programmable deformation, adaptive stiffness, and pattern transformation.

3.3.1.1 Euler Buckling Theory

Euler buckling theory describes the instability of slender columns subjected to compressive axial loads. According to this theory, when the compressive force acting on a column exceeds a certain critical value, the column becomes unstable and deflects laterally. The buckling behavior depends on the column's length, material stiffness, cross-sectional geometry, and boundary conditions. Euler's theory assumes that the column is perfectly straight, homogeneous, and subjected to purely axial loading. Under these assumptions, the critical buckling condition occurs when the compressive load reaches a threshold value at which the column can no longer maintain its straight configuration.

3.3.1.2 Critical Load Derivation

The critical load for buckling of a slender column can be derived from Euler's stability analysis and is expressed as:

$$P_{cr} = \frac{\pi^2 EI}{(KL)^2}$$

where:

P_{cr} = Critical buckling load
E = Young's modulus of the material
I = second moment of area of the column cross-section
L = length of the column
K = effective length factor depending on boundary conditions.

This equation shows that the buckling load increases with higher material stiffness and larger cross-sectional moment of inertia, while it decreases with increasing column length. Different boundary conditions such as pinned-pinned, fixed-fixed, fixed-free, or fixed-pinned affect the effective length factor and therefore the critical load.

3.3.1.3 Post-Buckling Behavior

After the critical load is exceeded, the structure enters the post-buckling regime, where significant lateral deflections occur. In many conventional structures, post-buckling deformation leads to structural failure. However, in mechanical metamaterials, the post-buckling behavior can be harnessed as a functional mechanism. Controlled buckling of slender elements within a lattice can produce large and reversible deformations, pattern transformations, and multistable configurations. This controlled instability enables metamaterials to exhibit tunable stiffness, energy absorption capability, and adaptive structural response under external loads.

3.3.2 Buckling as a Design Tool

In conventional structural engineering, buckling is typically regarded as an undesirable instability that may lead to structural failure. However, in mechanical metamaterials, buckling is intentionally utilized as a design mechanism to achieve novel and programmable mechanical responses. By carefully controlling geometry, boundary conditions, and material properties, engineers can exploit elastic instability to create structures capable of large reversible deformations, adaptive stiffness, and dynamic pattern transformations. These instability-driven mechanisms allow metamaterials to exhibit mechanical behaviors that are not achievable in traditional materials.

3.3.2.1 Instability-Induced Pattern Transformation

One of the most significant phenomena associated with buckling-driven metamaterials is pattern transformation. In periodic lattice structures composed of slender beams or cellular elements, compressive loading can trigger a coordinated buckling of the structural members. As a result, the original geometry of the unit cells transforms into a new configuration with a different structural pattern. This transformation may lead to changes in effective mechanical properties such as stiffness, Poisson's ratio, and anisotropy. Pattern transformation is widely observed in elastomeric metamaterials and cellular lattices, where structural symmetry can change under loading, enabling programmable mechanical responses.

3.3.2.2 Snap-Through and Snap-Back Behavior

Another important instability mechanism is snap-through behavior, which occurs when a structure rapidly transitions between two equilibrium configurations once a critical load or displacement is reached. This sudden change in configuration is often accompanied by a rapid release or absorption of stored elastic energy. The reverse transition, known as snap-back, occurs when the load is reduced and the structure returns to its original configuration. These mechanisms are frequently observed in curved beams, shell structures, and bistable metamaterial elements. Snap-through instabilities are particularly useful for creating fast mechanical switches, actuators, and energy-dissipating systems.

3.3.2.3 Bistable and Multistable Architectures

Buckling-driven mechanisms can also lead to bistable or multistable structural configurations, where a structure can remain stable in two or more distinct equilibrium states without continuous external energy input. Bistable metamaterials can switch between configurations through mechanical loading, thermal activation, or other stimuli. Multistable architectures extend this concept by enabling multiple stable deformation states, allowing structures to store mechanical information or exhibit programmable mechanical responses. These properties are valuable for applications such as deployable structures, soft robotic systems, adaptive aerospace components, and energy-absorbing materials.

3.3.3 Programmable Instability

Programmable instability refers to the deliberate design and control of elastic instabilities within mechanical metamaterials to achieve predictable and tunable deformation behavior. Instead of avoiding instability, engineers intentionally incorporate structural elements that undergo controlled buckling, snap-through, or pattern transformation when subjected to specific loading conditions. By carefully designing the geometry, stiffness distribution, and boundary conditions of the unit cells, the onset and progression of instability can be programmed to produce desired mechanical responses.

In programmable metamaterials, instability is often triggered through geometric features such as slender beams, curved elements, hinged joints, or lattice architectures that are sensitive to compressive or shear loading. When the applied load reaches a predetermined threshold, these elements undergo controlled buckling or rotation, resulting in a significant change in the structural configuration. The deformation path can be precisely designed so that the structure transitions through specific configurations, enabling adaptive mechanical functionality.

Programmable instability enables metamaterials to exhibit switchable mechanical properties, such as tunable stiffness, variable Poisson's ratio, and reversible shape transformation. For example, periodic cellular structures may undergo pattern transformation under compression, changing from one symmetric configuration to another. These transformations can be reversible, allowing the structure to recover its original geometry once the load is removed.

Such behavior is particularly useful in adaptive engineering systems, including deployable aerospace structures, soft robotics, morphing surfaces, and vibration control devices. In these applications, programmable instability allows structures to respond intelligently to external stimuli, enabling mechanical adaptability and reconfigurability. By harnessing elastic instability in a controlled manner, programmable metamaterials expand the functional capabilities of architected materials beyond traditional structural design.

3.3.4 Applications

Buckling-driven and instability-based mechanical metamaterials have attracted significant attention due to their ability to produce large, reversible, and programmable deformations. By exploiting controlled elastic instability, these metamaterials can transition between different structural configurations and mechanical states. Such characteristics enable a wide range of advanced engineering applications where adaptability, reconfigurability, and lightweight performance are essential.

3.3.4.1 Adaptive Structures

Adaptive structures are systems capable of adjusting their shape, stiffness, or mechanical response in response to changing environmental conditions or external loads. Instability-driven metamaterials provide an effective platform for adaptive behavior because their internal architecture can reconfigure through controlled buckling or pattern transformation. For example, lattice structures designed with slender beams may alter their stiffness or geometry when subjected to compressive loads, allowing the structure to adapt to different loading conditions while maintaining structural integrity.

3.3.4.2 Deployable Systems

Deployable systems are structures designed to compactly fold or collapse during storage and expand when deployed. Instability-based metamaterials are particularly suitable for such applications because their multistable configurations allow them to switch between compact and expanded states. These systems are widely used in space structures, antennas, solar panels, and portable engineering devices, where efficient packaging and rapid deployment are essential.

3.3.4.3 Soft Robotics

Soft robotic systems require materials that can undergo large, flexible, and reversible deformations without mechanical failure. Buckling-driven metamaterials provide programmable deformation pathways that enable soft robotic components to bend, twist, or expand in a controlled manner. By designing specific lattice geometries and incorporating flexible materials, these metamaterials can function as actuators, grippers, or locomotion mechanisms in soft robotic devices.

3.3.4.4 Morphing Aerospace Skins

Morphing aerospace structures are designed to change their shape during operation to improve aerodynamic performance and efficiency. Instability-driven metamaterials enable such morphing behavior by allowing structural components to transition between different shapes through controlled buckling or snap-through mechanisms. These adaptive skins can modify wing curvature, surface contours, or structural stiffness in response to flight conditions, leading to improved aerodynamic efficiency and reduced fuel consumption.

3.4 NEGATIVE STIFFNESS AND DAMPING BEHAVIOR

Negative stiffness and enhanced damping behavior are distinctive mechanical characteristics observed in certain mechanical metamaterials. Unlike conventional materials, where stiffness is positive and the restoring force increases with displacement, systems exhibiting negative stiffness display a temporary reduction in resistance when subjected to deformation. In such systems, an increase in displacement may lead to a decrease in restoring force over a specific range of deformation. Although negative stiffness is inherently unstable in isolated systems, it can be stabilized within composite or architected structures, allowing it to be utilized effectively for advanced mechanical functionality.

Negative stiffness behavior often arises from pre-buckled elements, bistable structures, or phase-like transitions within metamaterial architectures. For example, slender beams or curved elements may initially resist deformation but undergo sudden structural reconfiguration once a critical load is reached. During this transition region, the effective stiffness becomes negative, enabling rapid displacement with minimal additional force. When integrated within a stable framework, such elements can significantly enhance the overall mechanical response of the structure (Antoniadis et al., 2015).

One of the major benefits of incorporating negative stiffness elements into metamaterials is the ability to achieve exceptionally high damping capacity. During deformation, the structural rearrangement associated with instability and snap-through mechanisms dissipates mechanical energy effectively. This leads to increased vibration attenuation and improved energy dissipation performance. As a result, metamaterials with negative stiffness elements are particularly effective in vibration isolation systems and shock absorption applications.

Negative stiffness-based metamaterials are also capable of achieving ultra-high effective stiffness and damping when combined with conventional positive stiffness materials, forming hybrid systems that exhibit superior mechanical performance compared to traditional materials alone. These characteristics make such metamaterials valuable in engineering applications such as seismic protection systems, vibration control devices, noise suppression structures, and high-performance damping materials.

3.4.1 Concept of Negative Stiffness

Negative stiffness refers to a mechanical phenomenon in which a structure or material exhibits a reduction in restoring force with increasing deformation, resulting in a negative slope in the force–displacement relationship over a certain range of deformation. In conventional materials, stiffness is positive, meaning that the restoring force increases proportionally with displacement according to Hooke's law. However, in systems exhibiting negative stiffness, the structure tends to move further away from its equilibrium position once deformation begins, indicating an inherently unstable mechanical state.

Negative stiffness behavior is often associated with elastic instability, buckling, or phase-like transitions in structural elements. For example, a curved beam, buckled column, or bistable structure may initially resist deformation but suddenly reconfigure when a critical load is reached. During this transition phase, the force required to continue deformation decreases even as displacement increases, producing a negative stiffness region in the load–displacement curve. Although this behavior is unstable when occurring alone, it can be stabilized when combined with surrounding structures or positive stiffness materials.

In mechanical metamaterials, negative stiffness is typically achieved through architectural design, such as incorporating pre-buckled elements, snap-through mechanisms, or bistable unit cells within a lattice structure. When integrated within a stable framework, these negative stiffness elements can significantly enhance the overall mechanical performance of the system. Such structures can exhibit exceptionally high effective stiffness, enhanced damping capability, and improved vibration isolation.

The concept of negative stiffness is particularly important in the development of advanced metamaterials designed for energy absorption, vibration control, and adaptive mechanical systems. By carefully engineering the structural geometry and deformation mechanisms, negative stiffness can be harnessed to create materials with highly tunable mechanical responses that are not achievable in conventional materials.

3.4.2 Realization Mechanisms

Negative stiffness behavior in mechanical metamaterials is achieved through carefully designed structural mechanisms that introduce controlled instability within the material architecture. Since negative stiffness is inherently unstable when occurring alone, it is typically realized by embedding instability-driven elements within a stable framework or composite system. These mechanisms rely on structural features such as pre-buckled elements, curved beams, bistable units, and internal inclusions that undergo sudden structural transitions when subjected to external loads.

One common realization mechanism involves pre-buckled structural elements, such as curved beams or slender columns. These elements are initially designed in a slightly buckled configuration and can undergo further deformation when subjected to compressive or tensile loading. During this transition, the restoring force may temporarily decrease as displacement increases, producing a region of negative stiffness in the force–displacement response.

Another important mechanism is based on bistable structures, which possess two stable equilibrium configurations. When an external force is applied, the structure transitions between these configurations through a snap-through instability. During the transition phase, the system exhibits negative stiffness behavior as it moves between the two stable states. This mechanism is widely used in metamaterials designed for mechanical switching, energy absorption, and adaptive structural systems.

Negative stiffness can also be realized through embedded inclusions or composite architectures, where internal elements with instability-driven behavior are incorporated within a matrix of conventional positive stiffness materials. These inclusions may consist of buckled beams, rotating units, or hinged mechanisms that generate localized negative stiffness. When combined with the surrounding structure, the overall system remains stable while benefiting from enhanced damping and vibration isolation properties.

By integrating these realization mechanisms into metamaterial architectures, engineers can design structures that exhibit enhanced vibration damping, improved energy dissipation, and tunable mechanical response. These mechanisms are fundamental to the development of advanced metamaterials used in vibration control, seismic protection systems, adaptive structures, and impact-resistant engineering components.

3.4.3 Enhanced Damping Mechanisms

Enhanced damping mechanisms in mechanical metamaterials arise from the ability of their engineered architectures to dissipate mechanical energy more effectively than conventional materials. Damping refers to the process through which mechanical vibrations or dynamic loads are reduced by converting mechanical energy into other forms, such as heat. In metamaterials incorporating negative stiffness elements or instability-driven mechanisms, energy dissipation can occur through structural reconfiguration, internal friction, and hysteretic deformation, leading to significantly improved damping performance.

One of the primary sources of enhanced damping is the snap-through instability and structural transition observed in bistable or multistable metamaterial architectures. During deformation, these structures may undergo rapid configuration changes that absorb and dissipate a considerable amount of mechanical energy. The repeated transition between stable states produces hysteresis in the force–displacement response, which contributes to effective vibration attenuation.

Another important mechanism involves viscoelastic deformation and internal friction within the structural elements of the metamaterial. When combined with negative stiffness components, these effects can amplify energy dissipation by increasing the deformation pathways through which energy is absorbed. Additionally, metamaterial structures designed with buckled beams, rotating units, or flexible ligaments can undergo large reversible deformations that redistribute stresses and convert dynamic energy into structural motion.

Enhanced damping metamaterials are particularly valuable in applications requiring vibration isolation, shock absorption, and noise reduction. For instance, in seismic engineering, negative stiffness-based metamaterials can significantly reduce structural vibrations during earthquakes. Similarly, in aerospace and automotive systems, these materials can improve dynamic stability by suppressing unwanted vibrations.

3.4.4 Applications

Mechanical metamaterials exhibiting negative stiffness and enhanced damping behavior have significant applications in engineering systems where vibration control, energy dissipation, and dynamic stability are critical. By integrating instability-driven mechanisms and energy-dissipating architectures, these metamaterials can effectively reduce mechanical vibrations, absorb shocks, and suppress unwanted dynamic responses.

- **Vibration Isolation Systems**: One of the most important applications of negative stiffness metamaterials is in vibration isolation systems. Structures incorporating negative stiffness elements can achieve extremely high vibration attenuation by counteracting external dynamic forces. When combined with conventional positive stiffness components, these metamaterials can create systems with very low effective stiffness, allowing them to isolate sensitive equipment from vibrations. Such systems are widely used in precision instruments, aerospace structures, and high-performance machinery where vibration control is essential for operational accuracy and reliability.
- **Seismic Protection**: Negative stiffness metamaterials also play a significant role in seismic protection systems for buildings and infrastructure. During earthquakes, structures are subjected to large dynamic loads and ground vibrations. Metamaterials designed with instability-driven damping

mechanisms can absorb and dissipate seismic energy, thereby reducing the amplitude of vibrations transmitted to the structure. These systems can be incorporated into base isolators or structural dampers, improving the resilience of buildings and bridges against earthquake-induced damage.

- **Noise Suppression Structures**: Another important application is in noise suppression and acoustic vibration control. Mechanical metamaterials with enhanced damping properties can reduce the transmission of vibrational energy that generates noise. By absorbing and dissipating vibrational energy, these materials help minimize unwanted sound propagation in mechanical systems. This capability is particularly useful in automotive components, aerospace structures, industrial machinery, and architectural acoustics, where reducing vibration-induced noise is essential for improving performance and comfort.

3.5 TUNABLE STIFFNESS AND SHAPE MORPHING

Tunable stiffness and shape morphing are important functional characteristics of mechanical metamaterials that enable structures to adapt their mechanical properties and geometry in response to external stimuli or applied loads. Unlike conventional materials with fixed stiffness and shape, metamaterials with engineered architectures can dynamically modify their mechanical response through structural reconfiguration, internal mechanisms, or integration with responsive materials. These capabilities allow the development of adaptive systems capable of adjusting their performance according to changing operational requirements.

Tunable stiffness refers to the ability of a structure to vary its effective stiffness over a controlled range. This behavior can be achieved through geometric mechanisms such as rotating units, buckling elements, and multistable architectures. When external forces are applied, these structural elements can reorient, bend, or snap into different configurations, altering the load-bearing capacity and stiffness of the system. In some cases, tunable stiffness can also be controlled through external stimuli such as temperature, magnetic fields, or electrical signals when smart materials are incorporated into the metamaterial architecture.

Shape morphing involves controlled transformation of structural geometry without causing permanent damage to the material. This phenomenon is often achieved through structural mechanisms inspired by origami and kirigami designs, where folding patterns allow large geometric transformations while maintaining structural integrity. Differential strain mechanisms, multilayer structures, and programmable lattice geometries can also produce shape morphing behavior when subjected to mechanical, thermal, or electromagnetic stimuli.

The combination of tunable stiffness and shape morphing enables mechanical metamaterials to perform adaptive functions in various engineering applications. For instance, soft robotic systems require materials that can change stiffness and shape to perform gripping or locomotion tasks. In aerospace engineering, morphing structures can adjust aerodynamic surfaces during flight to optimize performance and reduce drag. Similarly, biomedical devices and adaptive architectural structures can benefit from materials capable of controlled deformation and structural reconfiguration.

3.5.1 Mechanisms of Tunable Stiffness

Tunable stiffness in mechanical metamaterials refers to the capability of a structure to modify its effective stiffness in response to external loading conditions or environmental stimuli. Unlike conventional materials with fixed mechanical properties, metamaterials achieve tunable stiffness through carefully engineered internal architectures that allow structural elements to reconfigure or deform in a controlled manner. These mechanisms enable structures to transition between soft and stiff states, providing adaptability and improved functionality in various engineering systems.

One common mechanism for achieving tunable stiffness is geometric reconfiguration. In architected lattice structures, structural elements such as beams, hinges, or rotating units can change orientation when subjected to external forces. As the geometry of the unit cells changes, the load distribution within the structure is altered, leading to variations in effective stiffness. For example, rotating unit structures or re-entrant lattices can transition between different configurations that exhibit distinct stiffness characteristics.

Another important mechanism involves elastic instability and buckling-driven behavior. Slender structural elements within the metamaterial may undergo controlled buckling when a critical load is reached. This buckling can cause a sudden change in deformation mode, leading to a significant modification of the structural stiffness. By designing the geometry and boundary conditions of these elements, engineers can control the onset and progression of buckling to achieve desired stiffness variations.

Tunable stiffness can also be achieved through the integration of smart or responsive materials, such as shape memory alloys, magnetorheological materials, or piezoelectric elements. These materials respond to external stimuli such as temperature, magnetic fields, or electric fields, enabling active control over the stiffness of the metamaterial structure. When incorporated into architected lattices or composite systems, these materials provide additional functionality by allowing real-time stiffness modulation.

3.5.2 Shape-Morphing Architectures

Shape morphing architectures refer to structural designs in mechanical metamaterials that enable controlled and reversible changes in shape or geometry in response to external stimuli such as mechanical loading, temperature, magnetic fields, or electric fields. Unlike conventional materials that maintain a fixed geometry under normal operating conditions, morphing metamaterials are engineered to undergo large and predictable deformations without compromising structural integrity. This capability is achieved through carefully designed internal architectures that allow structural elements to bend, rotate, fold, or expand during deformation.

One common approach to shape morphing involves origami- and kirigami-inspired architectures, where folding patterns and strategically placed cuts enable complex geometric transformations. Origami-based metamaterials rely on folding mechanisms to produce large shape changes from compact configurations, while kirigami designs incorporate cuts or slits that allow stretching and expansion. These architectures can transition between multiple geometric states, making them suitable for deployable and adaptive structures.

Another important mechanism is based on differential strain within multilayer structures. In such systems, layers composed of different materials or structural stiffness expand or contract at different rates when exposed to external stimuli. This mismatch in deformation produces bending or curvature, allowing the structure to morph into predetermined shapes. Such mechanisms are commonly used in soft robotics, flexible electronics, and adaptive surfaces.

Shape morphing can also be achieved through programmable lattice structures, where the arrangement and connectivity of unit cells are designed to produce coordinated deformation patterns. When mechanical loads or environmental stimuli are applied, the lattice elements reconfigure in a predictable manner, enabling complex three-dimensional shape transformations.

These morphing architectures are particularly valuable in advanced engineering applications such as soft robotic actuators, morphing aerospace wings, deployable space structures, biomedical devices, and adaptive architectural systems. By integrating geometric design with responsive materials and controlled deformation mechanisms, shape morphing metamaterials provide a powerful platform for developing structures capable of dynamic and multifunctional performance.

Figure 3.3 illustrates three fundamental architectural strategies used to achieve shape morphing behavior in mechanical metamaterials. The figure is divided into three sections representing different mechanisms: origami-based mechanisms, programmable lattice structures, and multilayer differential strain systems. In the first section, the origami mechanism demonstrates how a folded structure can transform from a compact folded configuration into a three-dimensional expanded geometry through controlled folding and unfolding. This mechanism enables large shape transformations while maintaining structural integrity. The second section presents a programmable lattice architecture, where a periodic network of interconnected structural elements transitions from an initial flat configuration to a curved or dome-like structure when mechanical loads are applied. The deformation occurs due to coordinated rotation and bending of the lattice members, allowing the structure to achieve

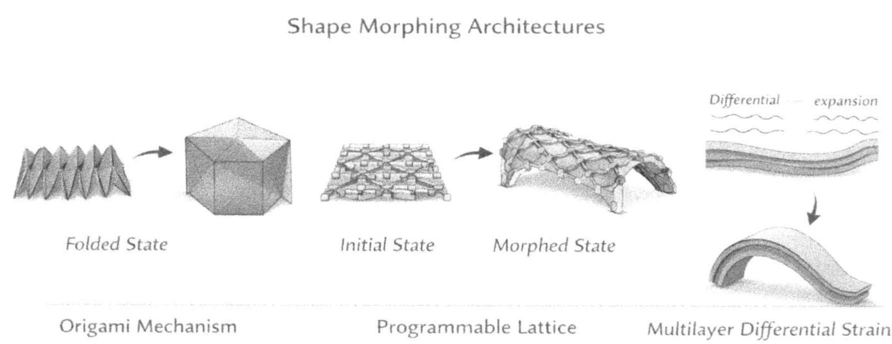

FIGURE 3.3 Shape morphing architectures in mechanical metamaterials.

controlled morphing. The third section illustrates multilayer differential strain, where layers composed of different materials or stiffness properties deform at different rates when subjected to external stimuli. This mismatch in strain produces bending or curvature, resulting in a morphing structure that changes from a straight configuration to a curved shape.

3.5.3 Stimuli-Responsive Metamaterials

Stimuli-responsive metamaterials are advanced mechanical metamaterials that can change their mechanical properties, geometry, or functional behavior in response to external stimuli such as temperature, magnetic fields, electric fields, light, or mechanical forces. Unlike passive materials with fixed properties, these metamaterials incorporate responsive materials or adaptive structural mechanisms that enable dynamic control over stiffness, shape, and deformation patterns. As a result, they can adapt to varying environmental conditions and perform programmable mechanical functions.

The responsiveness of these metamaterials is typically achieved by integrating smart materials into the metamaterial architecture. For example, shape memory alloys (SMAs) can recover their original shape when heated, allowing structures to change configuration through thermal activation. Similarly, magnetorheological materials respond to magnetic fields by altering their stiffness or damping characteristics, enabling active control of mechanical behavior. Piezoelectric materials can generate mechanical deformation when subjected to electric fields, providing precise actuation and tunable structural response.

Another approach to achieving stimulus-responsive behavior involves structural design and geometric reconfiguration. Architected lattice structures can be designed so that small environmental changes trigger large structural transformations. For instance, differential thermal expansion in multilayer materials can cause bending or curvature, enabling controlled shape morphing. Similarly, light-responsive polymers and hydrogels can swell or shrink under specific conditions, leading to adaptive deformation.

Stimuli-responsive metamaterials have numerous potential applications in soft robotics, biomedical devices, adaptive aerospace structures, sensors, and smart architectural systems. In soft robotics, these materials enable flexible actuators capable of complex movements. In aerospace engineering, they can be used to develop morphing surfaces that adjust aerodynamic properties during flight. Overall, stimuli-responsive metamaterials demonstrate how the integration of smart materials and architected structures can produce systems capable of intelligent, adaptive mechanical behavior.

3.5.4 Applications

Mechanical metamaterials with tunable stiffness and shape morphing capabilities have enabled the development of adaptive engineering systems that can respond dynamically to changing environmental conditions and functional requirements. These materials utilize programmable architectures, smart materials, and controlled deformation mechanisms to perform complex tasks that are difficult to achieve with conventional materials.

- **Soft Robotics**: One of the most promising applications of morphing metamaterials is in soft robotics, where flexibility and large reversible deformation are essential. Shape morphing structures allow robotic components such as grippers, actuators, and locomotion systems to bend, stretch, and adapt to different objects or environments. Tunable stiffness further enhances robotic performance by allowing the system to switch between soft and rigid states, improving both adaptability and load-bearing capability.
- **Deployable Aerospace Structures**: Morphing metamaterials are also highly valuable in deployable aerospace structures such as satellite antennas, solar panels, and space exploration devices. These systems must be compact during launch and expand or unfold once deployed in space. Metamaterials with programmable folding mechanisms and multistable architectures enable efficient packaging and reliable deployment while maintaining structural stability.
- **Adaptive Biomedical Devices**: In biomedical engineering, shape morphing metamaterials can be used to develop adaptive medical devices such as stents, implants, and prosthetic components. These devices can adjust their geometry to conform to complex biological structures, improving comfort, compatibility, and performance. Stimuli-responsive metamaterials can also enable medical devices that respond to temperature or physiological conditions within the body.
- **Smart Architectural Façades**: Another emerging application is in smart architectural façades, where buildings incorporate adaptive structures capable of responding to environmental changes. Morphing metamaterial panels can adjust their shape or stiffness to control sunlight, airflow, and thermal regulation within buildings. Such systems contribute to improved energy efficiency and enhanced environmental sustainability in modern architectural design.

3.6 IMPACT RESISTANCE, ENERGY ABSORPTION, AND LIGHT WEIGHTING

Impact resistance, energy absorption, and lightweight structural efficiency are among the most significant functional advantages offered by mechanical metamaterials. Unlike conventional materials, whose mechanical response is determined primarily by intrinsic material properties, metamaterials derive their performance from engineered architectures such as lattice structures, cellular networks, and hierarchical geometries. These architectures enable the design of structures that can withstand dynamic loads, dissipate impact energy effectively, and maintain high structural strength while minimizing weight.

One of the key advantages of mechanical metamaterials is their high impact resistance. When subjected to sudden loads or collisions, architected structures can distribute stresses more uniformly throughout the structure. The presence of interconnected beams, nodes, and cellular elements allows the structure to deform progressively rather than failing abruptly. This progressive deformation mechanism reduces stress concentration and enhances the ability of the structure to resist impact damage.

Another important characteristic is enhanced energy absorption capability. Mechanical metamaterials often employ mechanisms such as buckling, bending, and controlled collapse of lattice elements to dissipate mechanical energy. During impact or compression, these structural elements undergo sequential deformation and structural reconfiguration, converting kinetic energy into internal deformation energy. This property is particularly valuable in applications where shock mitigation and crash protection are required.

Lightweighting is another critical benefit achieved through architected metamaterial designs. By replacing solid materials with periodic lattice or cellular structures, engineers can significantly reduce structural mass while maintaining sufficient mechanical strength and stiffness. These lightweight structures exhibit high specific strength and stiffness, meaning they provide excellent mechanical performance relative to their weight. This characteristic is especially important in industries such as aerospace, automotive engineering, and transportation, where reducing weight directly contributes to improved efficiency and performance.

3.6.1 Fundamentals of Impact Mechanics

Impact mechanics deals with the behavior of materials and structures when subjected to sudden dynamic loads or high-speed collisions. Unlike static loading conditions where forces are applied gradually, impact loading occurs over a very short time interval and involves rapid transfer of momentum and energy. In mechanical metamaterials, understanding impact mechanics is essential for designing structures capable of absorbing energy, resisting damage, and mitigating shock loads.

When an impact event occurs, a large amount of kinetic energy from the impacting body is transferred to the structure. This energy is distributed through the material in the form of stress waves that propagate through the structure. The magnitude and speed of these stress waves depend on material properties such as density, elastic modulus, and internal architecture. In metamaterial structures, engineered lattice geometries can alter the propagation of these waves, helping to reduce peak stresses and distribute loads more uniformly.

Another important concept in impact mechanics is strain-rate sensitivity, which describes how material behavior changes when deformation occurs at high speeds. Many materials exhibit increased stiffness and strength under high strain-rate conditions. Mechanical metamaterials can take advantage of this behavior through their structural design, enabling them to withstand dynamic loads more effectively than conventional materials.

Impact response also involves mechanisms such as progressive collapse, buckling of structural members, and plastic deformation, which contribute to energy dissipation. In cellular or lattice-based metamaterials, the controlled collapse of structural elements during impact allows the structure to absorb significant amounts of energy while preventing catastrophic failure.

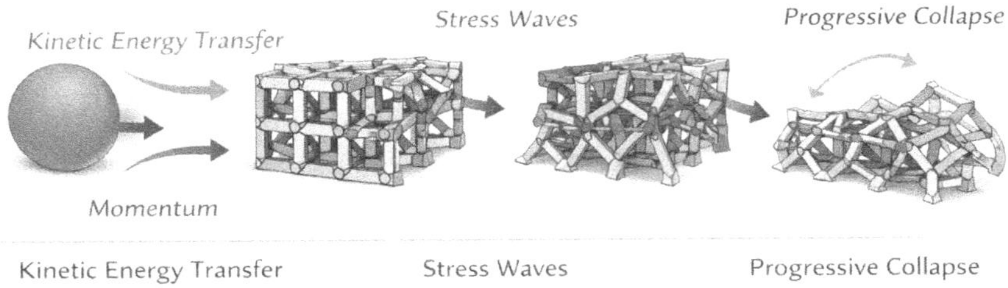

FIGURE 3.4 Fundamentals of impact mechanics in mechanical metamaterials.

Understanding these fundamental principles of impact mechanics enables engineers to design metamaterial structures with improved impact resistance, shock absorption capability, and structural safety, making them suitable for applications in automotive crash protection, aerospace structures, protective equipment, and energy-absorbing devices.

Figure 3.4 illustrates the fundamental processes that occur when a mechanical metamaterial structure is subjected to impact loading. The figure is divided into three stages: kinetic energy transfer, stress wave propagation, and progressive collapse. In the first stage, a moving object (represented by a sphere) strikes the metamaterial structure, transferring kinetic energy and momentum to the structure. This sudden transfer of energy initiates a dynamic response within the material. In the second stage, the impact generates stress waves that propagate through the internal lattice architecture of the metamaterial. These stress waves distribute the impact forces across the structural network, reducing localized stress concentration. In the final stage, the metamaterial undergoes progressive collapse, where individual lattice members bend, buckle, or deform sequentially. This controlled deformation mechanism allows the structure to absorb and dissipate a significant portion of the impact energy while preventing catastrophic failure.

3.6.2 Energy Absorption Mechanisms

Energy absorption is one of the most important functional characteristics of mechanical metamaterials, particularly in applications involving impact protection, crashworthiness, and shock mitigation. Unlike conventional solid materials, metamaterials are designed with cellular and lattice architectures that enable controlled deformation and progressive collapse under dynamic loading. These structural mechanisms allow the material to convert kinetic energy into internal deformation energy, thereby reducing the magnitude of forces transmitted to surrounding structures.

One of the primary mechanisms responsible for energy absorption in mechanical metamaterials is buckling of structural members. Slender beams and ligaments within the lattice architecture can buckle under compressive loading, allowing the structure to deform without immediate failure. This controlled buckling distributes the applied load across the structure and helps dissipate energy through elastic and plastic deformation.

Another important mechanism is plastic deformation of structural elements. When the applied stress exceeds the elastic limit of the material, permanent deformation occurs within the lattice members. This process consumes a significant amount of mechanical energy and contributes to the overall energy absorption capability of the metamaterial. The presence of interconnected unit cells allows this deformation to occur progressively throughout the structure.

Mechanical metamaterials also exhibit progressive collapse of cellular structures, where individual unit cells collapse sequentially under compression or impact. This sequential deformation mechanism prevents sudden catastrophic failure and enables the structure to absorb energy gradually over an extended deformation range. Additionally, mechanisms such as rotation of structural units, bending of ligaments, and friction between structural elements further enhance energy dissipation.

Through careful control of unit-cell geometry, topology, and material properties, engineers can design metamaterials with optimized energy absorption characteristics. These structures are widely used in automotive crash structures, protective helmets, aerospace components, and energy-absorbing barriers, where efficient shock mitigation and structural safety are critical.

Figure 3.5 illustrates the primary mechanisms through which mechanical metamaterials absorb and dissipate energy when subjected to compressive or impact loading. The figure highlights three important deformation mechanisms: buckling, plastic deformation, and progressive collapse. In the first mechanism, slender structural members within the lattice architecture undergo elastic buckling when compressive forces are applied. This buckling allows the structure to deform without immediate failure while distributing the applied load throughout the lattice network. The second mechanism demonstrates plastic deformation, where the stress within the structural elements exceeds the elastic limit of the material. As a result, permanent deformation occurs in certain lattice members, converting mechanical energy into internal deformation energy and heat, thereby increasing the overall energy absorption capability. The third mechanism represents progressive collapse, where individual unit cells of the lattice structure collapse sequentially under increasing load. This controlled and gradual collapse prevents catastrophic failure and allows the structure to absorb energy over a larger deformation range. Together, these mechanisms illustrate how

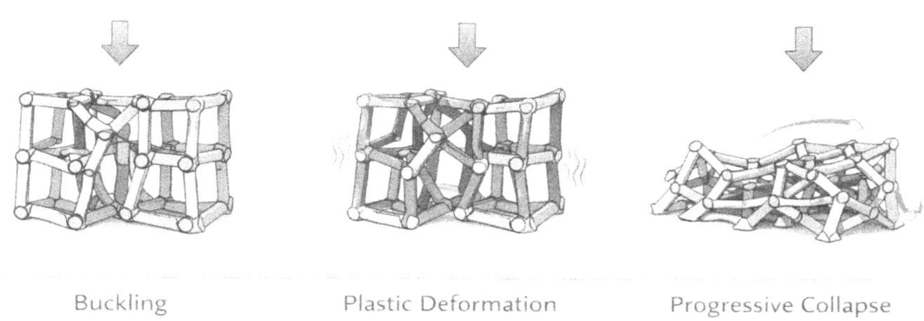

FIGURE 3.5 Energy absorption mechanisms in mechanical metamaterials.

engineered cellular architectures enable mechanical metamaterials to achieve superior impact resistance, shock mitigation, and energy dissipation, making them highly suitable for applications such as crashworthy structures, protective equipment, and aerospace energy-absorbing components.

3.6.3 Lightweight Structural Efficiency

Lightweight structural efficiency is one of the most significant advantages of mechanical metamaterials, enabling the development of structures that provide high mechanical performance while minimizing material usage and overall weight. In conventional solid materials, increasing strength and stiffness generally requires adding more material, which leads to increased weight. In contrast, mechanical metamaterials achieve superior performance through engineered lattice architectures and cellular structures, where the arrangement of structural elements rather than material quantity determines the overall mechanical behavior.

Mechanical metamaterials often utilize periodic lattice networks, honeycomb structures, and hierarchical architectures to create lightweight frameworks with high load-bearing capacity. These structures distribute loads through interconnected beams and nodes, allowing the structure to maintain mechanical stability while significantly reducing material volume. As a result, metamaterials can achieve high specific strength and specific stiffness, which are measures of mechanical performance relative to weight.

Another important aspect of lightweight structural efficiency is the ability of metamaterial architectures to maintain structural integrity even with a large fraction of void space. The presence of these voids reduces overall density without compromising the load-carrying capability of the structure. Additionally, hierarchical and multiscale lattice designs can further improve mechanical performance by optimizing load transfer across different structural scales.

Lightweight metamaterial structures are particularly valuable in industries where weight reduction directly improves efficiency and performance. In aerospace engineering, reducing structural weight leads to improved fuel efficiency and payload capacity. In the automotive industry, lightweight crash structures enhance vehicle safety while improving energy efficiency. Similarly, lightweight metamaterials are used in civil engineering, transportation systems, and protective equipment, where high strength-to-weight ratios are essential.

Figure 3.6 illustrates the concept of lightweight structural efficiency by comparing three types of structural configurations: solid material, lattice metamaterial, and honeycomb metamaterial. The first configuration represents a conventional solid material block, which contains a large amount of material and therefore exhibits a high weight-to-strength ratio. Although it provides structural strength, the excessive material usage results in increased weight, making it less efficient for applications where weight reduction is critical. The second configuration shows a lattice metamaterial structure, composed of interconnected beams and nodes that create a periodic cellular architecture. This structure significantly reduces the amount of material used while maintaining load-bearing capability, resulting in a low weight-to-strength ratio and improved structural efficiency. The third configuration presents a honeycomb

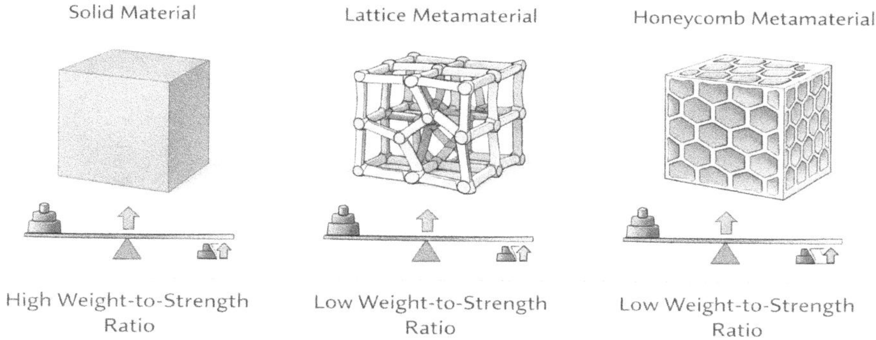

FIGURE 3.6 Comparison of lightweight structural architectures in mechanical metamaterials.

metamaterial, which uses hexagonal cellular geometry to distribute loads effectively while minimizing material consumption. Honeycomb structures are widely known for their excellent strength-to-weight performance and structural stability. The balance diagrams below each structure conceptually illustrate the relationship between weight and mechanical performance.

Table 3.1 summarizes the comparison of three structural configurations. It highlights how lattice and honeycomb metamaterial architectures achieve superior lightweight structural efficiency compared with conventional solid materials by reducing material usage while maintaining mechanical performance.

3.6.4 Crashworthiness and Protective Systems

Crashworthiness refers to the ability of a structure or material to absorb impact energy and protect occupants or critical components during collisions or sudden loading events. Mechanical metamaterials have gained significant attention in crashworthy and protective systems due to their engineered cellular architectures, which enable controlled deformation and efficient energy dissipation. Unlike conventional solid materials that may fail abruptly under impact, metamaterial structures are designed to deform progressively, thereby reducing the magnitude of transmitted forces and enhancing safety.

One of the key advantages of metamaterials in crashworthiness applications is their ability to undergo progressive collapse mechanisms. During impact or compression, the structural elements within lattice or cellular architectures deform sequentially through buckling, bending, and plastic deformation. This controlled collapse absorbs a large portion of the kinetic energy generated during a collision, preventing sudden structural failure and reducing damage to surrounding components.

Mechanical metamaterials also offer significant benefits in protective systems, where impact mitigation and energy absorption are critical. Structures such as auxetic lattices, hierarchical cellular networks, and honeycomb metamaterials can distribute stresses uniformly and prevent localized stress concentration. These features enhance the ability of protective materials to withstand high-energy impacts while maintaining structural integrity.

TABLE 3.1
Comparison of Lightweight Structural Efficiency in Different Architectures

Structure Type	Structural Description	Material Usage	Weight-to-Strength Ratio	Key Mechanical Characteristics	Typical Applications
Solid Material	A fully dense structure with no internal voids or cellular architecture.	High material consumption	High	Provides high stiffness and strength but results in increased structural weight and inefficient material usage.	Conventional machine components, solid structural blocks
Lattice Metamaterial	A periodic network of interconnected beams and nodes forming an open cellular structure.	Moderate to low material usage	Low	High specific strength and stiffness, efficient load distribution, and significant weight reduction.	Aerospace structures, lightweight automotive components, advanced engineering systems
Honeycomb Metamaterial	A cellular structure composed of repeating hexagonal cells that distribute loads efficiently.	Low material consumption	Low	Excellent strength-to-weight ratio, high structural stability, and good energy absorption characteristics.	Aircraft panels, sandwich structures, protective components

Due to their superior energy absorption and lightweight characteristics, metamaterials are increasingly used in applications such as automotive crash structures, protective helmets, body armor, aerospace impact protection systems, and blast-resistant structures. In automotive engineering, metamaterial-based crash absorbers can reduce the forces transmitted to passengers during collisions. In aerospace applications, lightweight impact-resistant structures help protect aircraft components while minimizing structural weight.

Figure 3.7 illustrates several practical applications of mechanical metamaterials in crashworthiness and protective systems, highlighting how architected cellular structures improve impact protection and energy absorption. The figure presents three key examples: automotive crash absorbers, protective helmet padding, and body armor systems. In the first example, a vehicle equipped with a metamaterial-based automotive crash absorber demonstrates how lattice structures positioned in the front impact zone can deform progressively during a collision. This controlled deformation absorbs a significant portion of the kinetic energy and reduces the force transmitted to the main vehicle structure. The second example shows protective helmet padding, where auxetic or cellular metamaterial layers inside the helmet distribute impact forces over a larger area and enhance shock absorption, thereby improving head protection. The third example illustrates body armor systems, where metamaterial panels embedded in protective vests help dissipate impact energy from projectiles or blunt forces, increasing safety while maintaining lightweight structural efficiency. The balance diagrams below each illustration conceptually indicate the improved strength-to-weight performance achieved through metamaterial architectures.

Table 3.2 summarizes the major crashworthiness and protective applications of mechanical metamaterials. The table highlights how cellular and lattice architectures enhance energy absorption, impact resistance, and lightweight structural performance across automotive, protective equipment, and defense applications.

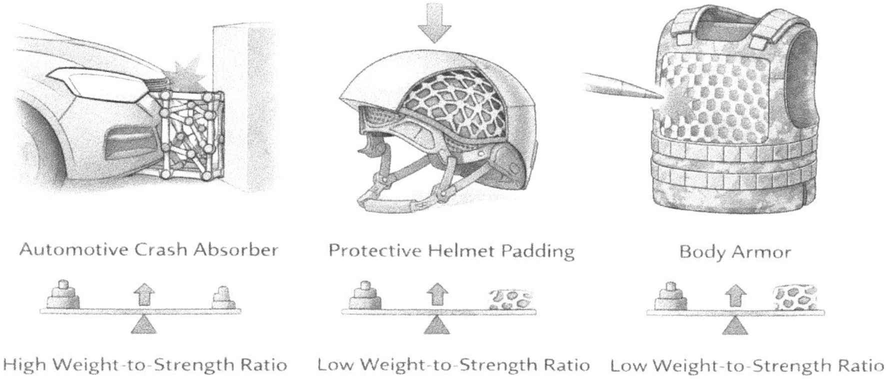

FIGURE 3.7 Crashworthiness and protective applications of mechanical metamaterials.

TABLE 3.2
Applications of Mechanical Metamaterials in Crashworthiness and Protective Systems

Application	Structural Configuration	Working Mechanism	Key Advantages	Typical Use Cases
Automotive Crash Absorber	Lattice or cellular metamaterial structures integrated into vehicle impact zones.	During collision, the lattice elements undergo progressive buckling and controlled collapse, absorbing kinetic energy and reducing transmitted forces.	High energy absorption, improved crashworthiness, lightweight design.	Automotive bumpers, crumple zones, impact absorbers in vehicles.
Protective Helmet Padding	Auxetic or cellular metamaterial layers placed inside helmet structures.	Impact forces are distributed through the cellular architecture, allowing deformation and energy dissipation over a larger area.	Improved shock absorption, enhanced head protection, lightweight comfort.	Sports helmets, military helmets, industrial safety helmets.
Body Armor Systems	Metamaterial panels or layered cellular structures embedded in protective vests.	The lattice network absorbs and redistributes impact energy from projectiles or blunt forces through structural deformation.	High impact resistance, lightweight protection, enhanced structural durability.	Military body armor, personal protective equipment, law enforcement gear.

3.6.5 Performance Metrics

Performance metrics are essential for evaluating the effectiveness of mechanical metamaterials in impact resistance, energy absorption, and crashworthiness applications. These metrics provide quantitative measures that allow engineers to compare different structural designs and materials based on their ability to absorb energy, resist deformation, and maintain structural integrity under dynamic loading conditions. By analyzing these parameters, researchers can optimize metamaterial architectures for improved mechanical performance.

One of the most important performance metrics is specific energy absorption (SEA), which represents the amount of energy absorbed per unit mass of the structure. SEA is calculated by dividing the total absorbed energy by the mass of the structure. A higher SEA value indicates that the structure can absorb more energy while maintaining a lightweight design, making it highly desirable in applications such as automotive crash structures and aerospace components.

Another important parameter is crushing force efficiency (CFE), which measures how uniformly a structure absorbs energy during compression or impact. CFE is defined as the ratio of the mean crushing force to the maximum peak force experienced during deformation. A higher CFE value indicates smoother energy absorption and reduced force spikes, which improves structural stability and safety during impact events.

Energy absorption capacity is also commonly used to evaluate crashworthiness performance. This metric represents the total mechanical energy dissipated during deformation or collapse of the structure. It is typically obtained from the area under the force–displacement curve during compression testing.

Other related performance indicators include peak crushing force, which represents the maximum force experienced during impact, and stroke efficiency, which measures how effectively the deformation length is utilized for energy absorption. By optimizing these performance metrics through architectural design and material selection, mechanical metamaterials can achieve superior impact resistance and structural efficiency.

3.6.5.1 Specific Energy Absorption

SEA is one of the most important performance metrics used to evaluate the energy absorption capability of materials and structures during impact or compression. It represents the amount of mechanical energy absorbed per unit mass of a structure and is widely used to assess the efficiency of lightweight crashworthy systems. In mechanical metamaterials, SEA is particularly significant because these materials are often designed to achieve maximum energy absorption while maintaining minimal weight.

Mathematically, SEA is expressed as:

$$SEA = \frac{E_{absorbed}}{m}$$

where:

$E_{absorbed}$ = total energy absorbed during deformation or crushing

m = mass of the structure.

The absorbed energy is typically determined from the area under the force–displacement curve obtained during compression or impact testing. A larger area under this curve indicates greater energy absorption capacity.

Mechanical metamaterials with engineered lattice or cellular architectures often exhibit high SEA values because their structures deform progressively through mechanisms such as buckling, bending, and collapse of unit cells. This controlled deformation allows the structure to dissipate large amounts of energy without significant weight increase. Consequently, metamaterials with high SEA are highly desirable for applications requiring impact protection and crash energy management.

High SEA performance is particularly important in industries such as automotive engineering, aerospace structures, protective equipment, and defense systems, where lightweight structures must still provide effective energy absorption. By optimizing unit-cell geometry, material selection, and structural topology, engineers can design metamaterials with significantly improved SEA compared with conventional solid materials.

3.6.5.2 Crushing Force Efficiency (CFE)

CFE is an important performance metric used to evaluate the stability and efficiency of energy absorption during the crushing or compression of a structure. It indicates how uniformly a structure absorbs energy while undergoing deformation. In crashworthiness applications, maintaining a stable and controlled crushing force is essential because sudden peaks in force can transmit excessive loads to protected components or occupants.

CFE is defined as the ratio of the mean crushing force to the maximum peak crushing force experienced during deformation. It can be expressed mathematically as:

$$CFE = \frac{F_{mean}}{F_{peak}}$$

where:

F_{mean} = average force during the crushing process
F_{peak} = maximum force experienced at the beginning of deformation.

A higher CFE value indicates more stable and uniform energy absorption, which means the structure deforms progressively with minimal force fluctuations. Ideally, the crushing force should remain nearly constant throughout the deformation process, ensuring that impact loads are distributed evenly.

Mechanical metamaterials often exhibit improved CFE compared with conventional materials due to their cellular and lattice architectures. These structures deform through mechanisms such as buckling, bending, and progressive collapse of unit cells, which help maintain a relatively steady crushing force during compression.

As a result, metamaterials can absorb impact energy efficiently while minimizing dangerous force spikes.

High CFE values are particularly desirable in applications such as automotive crash structures, protective barriers, aerospace components, and impact mitigation systems, where controlled deformation and stable energy absorption are essential for improving safety and structural reliability.

3.6.5.3 Energy Absorption Capacity

Energy absorption capacity refers to the total amount of mechanical energy that a material or structure can absorb during deformation before failure or complete collapse. It is one of the most important performance indicators in evaluating the effectiveness of structures designed for impact protection, crashworthiness, and shock mitigation. In mechanical metamaterials, the engineered cellular or lattice architectures enable efficient energy absorption through controlled deformation mechanisms.

The energy absorption capacity is typically determined from the force–displacement curve obtained during compression or impact testing. Mathematically, it is expressed as the area under the force–displacement curve:

$$E_{abs} = \int F\,dx$$

where:

E_{abs} = energy absorbed during deformation
F = applied force
x = displacement during deformation.

A larger area under the force–displacement curve indicates that the structure can absorb a greater amount of energy. Mechanical metamaterials often exhibit high energy absorption capacity because their internal architectures allow progressive collapse of unit cells, buckling of structural members, bending of ligaments, and plastic deformation. These mechanisms dissipate mechanical energy gradually rather than causing sudden structural failure.

High energy absorption capacity is particularly important in engineering systems that require impact resistance and crash protection, such as automotive crumple zones, aerospace protective structures, helmets, body armor, and energy-absorbing barriers. By optimizing unit-cell geometry, material distribution, and structural topology, mechanical metamaterials can achieve significantly higher energy absorption performance compared with conventional solid materials while maintaining lightweight structural efficiency.

3.6.5.4 Peak Crushing Force

Peak crushing force (PCF) is an important performance metric used to evaluate the maximum force experienced by a structure during the initial stage of compression or impact loading. It represents the highest force recorded at the beginning of the deformation process before the structure starts to undergo progressive collapse. In crashworthiness and protective system design, controlling the PCF is crucial because excessive peak forces can transmit large loads to protected components or occupants, potentially causing damage or injury.

During an impact event, the structure initially resists deformation until the applied load reaches the PCF. Once this peak value is exceeded, structural elements such as beams, ligaments, or cellular units begin to buckle, bend, or collapse, initiating the energy absorption phase. In well-designed mechanical metamaterials, the structural architecture is optimized to reduce the magnitude of the PCF, allowing the structure to transition smoothly into a progressive collapse regime.

Lower PCF is desirable because it reduces the sudden transmission of high forces during impact. Metamaterials with engineered lattice architectures can achieve this by introducing mechanisms such as controlled buckling, hierarchical cellular structures, and graded density designs, which distribute loads more evenly across the structure. This results in a smoother force–displacement response and improved crashworthiness performance.

PCF is commonly evaluated from the force–displacement curve obtained during compression testing. It corresponds to the highest force value observed before significant structural collapse occurs. In applications such as automotive crash absorbers, aerospace protective components, and energy-absorbing barriers, minimizing PCF while maintaining high energy absorption capacity is essential for improving safety and structural efficiency.

3.7 CASE STUDY: AUTOMOTIVE CRASH ABSORBER USING MECHANICAL METAMATERIALS

3.7.1 Introduction

Automotive crash absorbers are critical safety components designed to mitigate the impact energy generated during vehicle collisions. Traditional crash absorbers are typically made from metallic crumple zones or foam-based structures that deform plastically to absorb kinetic energy. However, recent advances in mechanical metamaterials have enabled the development of lightweight crash absorbers with significantly improved energy absorption efficiency and controlled deformation mechanisms. By incorporating engineered lattice architectures, these metamaterial-based absorbers can provide superior crashworthiness while reducing structural weight.

3.7.2 Design Concept

In this case study, a lattice-based mechanical metamaterial crash absorber is integrated into the front bumper structure of a vehicle. The absorber consists of a periodic cellular lattice structure fabricated using lightweight metallic or polymeric materials through additive manufacturing techniques such as selective laser melting (SLM) or fused deposition modeling (FDM). The lattice architecture is designed to undergo progressive buckling and controlled collapse under impact loading, thereby absorbing a large portion of the collision energy.

The metamaterial unit cells are arranged in a hierarchical structure that distributes stresses throughout the absorber. This design ensures that deformation occurs gradually rather than abruptly, preventing sudden failure and reducing the peak force transmitted to the vehicle frame.

3.7.3 Working Mechanism

When a collision occurs, the following sequence of events takes place:

- Initial Impact: The moving vehicle transfers kinetic energy to the crash absorber.
- Stress Wave Propagation: Impact forces propagate through the lattice architecture.
- Progressive Collapse: Structural members within the lattice begin to buckle and deform sequentially.
- Energy Dissipation: The deformation of the lattice converts kinetic energy into internal deformation energy and heat.
- Load Reduction: The controlled collapse reduces the peak force transmitted to the passenger cabin.

3.7.4 Performance Metrics

The effectiveness of the metamaterial crash absorber is evaluated using several crashworthiness metrics:

- Specific energy absorption (SEA): High SEA indicates efficient energy absorption per unit mass.
- Crushing force efficiency (CFE): Ensures uniform deformation during impact.
- Peak crushing force (PCF): Lower PCF reduces the risk of structural damage.
- Energy absorption capacity: Total energy dissipated during deformation.

Experimental and simulation studies have shown that lattice-based metamaterial crash absorbers can increase energy absorption efficiency by up to 30%–50% compared to conventional structures, while also reducing overall structural weight.

Figure 3.8 illustrates the integration of a metamaterial-based crash absorber within the front bumper structure of a vehicle. The figure shows a vehicle approaching an obstacle, with the metamaterial lattice absorber positioned between the bumper and the vehicle frame. During impact, the lattice architecture undergoes controlled progressive collapse, where individual structural members bend and buckle sequentially. This deformation mechanism distributes impact forces across the structure and absorbs a significant portion of the collision energy. As a result, the force transmitted to the vehicle frame and passenger cabin is significantly reduced. The figure highlights how the engineered cellular architecture enhances crashworthiness while maintaining lightweight structural efficiency, making mechanical metamaterials highly suitable for modern automotive safety systems.

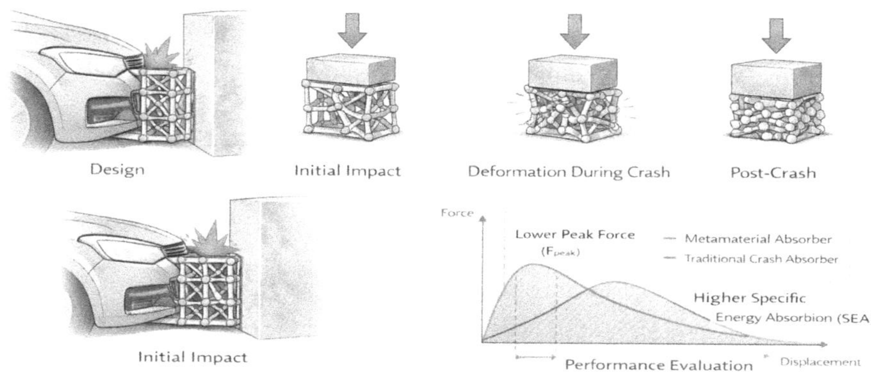

FIGURE 3.8 Lattice-based mechanical metamaterial crash absorber for automotive impact protection.

3.8 CASE STUDY: AUXETIC METAMATERIALS IN PROTECTIVE HELMETS

3.8.1 Introduction

Protective helmets are widely used in sports, military, construction, and transportation industries to reduce head injuries during impact events. Traditional helmet liners are commonly made of expanded polystyrene (EPS) foam, which absorbs impact energy through compression. However, these materials often suffer from limited energy absorption efficiency and poor repeated-impact performance. Auxetic mechanical metamaterials have emerged as a promising alternative due to their negative Poisson's ratio, which enhances impact resistance and energy distribution.

3.8.2 Design Concept

In this design, the helmet liner consists of a re-entrant auxetic lattice structure fabricated using polymer-based additive manufacturing techniques such as FDM or selective laser sintering (SLS). The auxetic unit cells expand laterally when compressed, which increases local density around the impact zone.

The liner is structured as a multilayer auxetic lattice that conforms to the internal geometry of the helmet shell. This design allows the helmet to maintain comfort while providing superior mechanical protection.

3.8.3 Working Mechanism

When an impact occurs:

1. Initial contact generates compressive forces on the helmet liner.
2. The auxetic unit cells expand laterally, increasing local stiffness.
3. The structure undergoes controlled deformation and energy dissipation.
4. Impact energy is distributed across a larger region, reducing localized stress.

3.8.4 Performance Benefits

Auxetic metamaterial liners provide several advantages:

- Improved impact resistance
- Higher energy absorption capacity
- Enhanced indentation resistance
- Better load distribution

3.8.5 Applications

- Sports helmets
- Military combat helmets
- Industrial safety helmets
- Bicycle and motorcycle helmets

Figure 3.9 illustrates the application of auxetic mechanical metamaterials in protective helmet systems and demonstrates how the unique structural geometry improves impact protection. The figure presents three main stages: design configuration, impact event, and impact distribution within the helmet liner. In the first stage, the helmet is shown with an internal auxetic lattice liner composed of re-entrant cellular structures. This engineered architecture forms the primary energy-absorbing layer inside the helmet shell. In the second stage, an external object strikes the helmet, representing an impact event that generates compressive forces on the auxetic liner. Due to the negative Poisson's ratio of the auxetic structure, the unit cells expand laterally during compression, increasing the local density and stiffness around the impact zone. In the third stage, the figure illustrates impact distribution, where the deformation of the auxetic lattice spreads the applied load across a larger area of the helmet liner. The inset diagram highlights the re-entrant auxetic geometry, which allows structural members to rotate and expand during deformation. This mechanism enhances energy absorption, reduces localized stress concentration, and improves overall impact resistance. As a result, auxetic metamaterial helmet liners provide higher energy absorption, improved indentation resistance, and better load distribution compared with conventional foam-based liners, making them highly effective for advanced protective helmet designs used in sports, military, and industrial safety equipment.

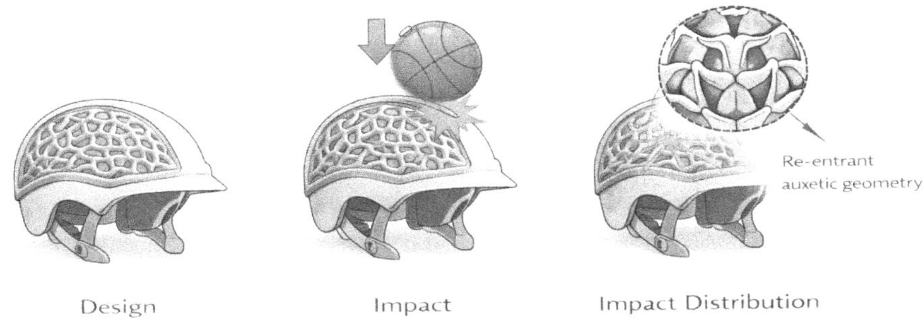

FIGURE 3.9 Impact protection mechanism of auxetic metamaterial helmet liners.

3.9 CASE STUDY: METAMATERIAL ENERGY-ABSORBING PANELS FOR AEROSPACE STRUCTURES

3.9.1 Introduction

Weight reduction is a critical requirement in aerospace engineering, as lower structural weight leads to improved fuel efficiency and payload capacity. However, aerospace structures must also withstand high impact loads, vibration, and dynamic stresses. Mechanical metamaterials with cellular architectures provide an effective solution by combining lightweight properties with high energy absorption capability.

3.9.2 Design Concept

In this case study, aerospace panels are designed using hierarchical lattice metamaterial cores sandwiched between thin composite skins. The lattice core is fabricated using SLM or electron beam melting (EBM) to produce high-strength metallic architectures.

The metamaterial core consists of octet-truss or tetrahedral lattice structures, which are known for their excellent stiffness-to-weight ratio.

3.9.3 Working Mechanism

During impact or dynamic loading:

1. Stress waves propagate through the sandwich panel.
2. The lattice core undergoes controlled deformation and buckling.
3. Impact energy is dissipated through bending and collapse of lattice members.
4. The outer skins remain protected from excessive damage.

3.9.4 Performance Advantages

Metamaterial aerospace panels provide:

- High specific stiffness
- Excellent impact resistance
- Superior energy absorption capability
- Significant weight reduction

3.9.5 Applications

- Aircraft fuselage panels
- Satellite structural components
- Spacecraft protective shielding
- Lightweight aerospace structural elements

3.10 CASE STUDY: SEISMIC METAMATERIAL BARRIERS FOR EARTHQUAKE PROTECTION

3.10.1 Introduction

Earthquakes generate strong ground vibrations that can cause severe damage to buildings and infrastructure. Conventional seismic protection methods include base isolators and damping systems, but recent research has introduced seismic metamaterials capable of manipulating and attenuating seismic waves.

3.10.2 Design Concept

Seismic metamaterials consist of periodic arrays of underground structural elements, such as cylindrical columns or lattice structures, arranged around buildings or critical infrastructure. These structures act as wave barriers that modify the propagation of seismic waves.

The metamaterial design creates bandgaps, which are frequency ranges where seismic waves cannot propagate effectively.

3.10.3 Working Mechanism

During an earthquake:

1. Seismic waves travel through the ground toward a building.
2. The metamaterial barrier interacts with the incoming waves.
3. Wave energy is reflected, scattered, or attenuated by the periodic structure.
4. The transmitted vibration energy reaching the structure is significantly reduced.

3.10.4 Performance Advantages

Seismic metamaterials provide:

- Reduced ground vibration transmission
- Improved structural stability
- Enhanced seismic resilience
- Passive vibration control without external power

3.10.5 Applications

- Earthquake-resistant buildings
- Bridge foundations
- Critical infrastructure protection
- Nuclear facility safety systems

3.11 SUMMARY

Mechanical metamaterials exhibit unique mechanical properties that arise primarily from their engineered architecture rather than intrinsic material composition. This chapter examined several key functional behaviors, including auxetic behavior with negative Poisson's ratio, elastic instability and buckling-driven mechanisms, negative stiffness and enhanced damping characteristics, tunable stiffness, and shape morphing architectures. These properties allow metamaterials to achieve unconventional responses such as controlled deformation, adaptive stiffness, programmable instability, and improved vibration isolation. The chapter also discussed the mechanisms responsible for impact resistance, energy absorption, and lightweight structural efficiency, which are critical for applications in crashworthiness and protective systems. Performance evaluation metrics such as SEA, CFE, PCF, and total energy absorption capacity were introduced to assess the effectiveness of metamaterial structures under dynamic loading. Finally, several practical case studies, including automotive crash absorbers, auxetic helmet liners, aerospace energy-absorbing panels, and seismic metamaterial barriers, demonstrated how these architected materials can significantly improve safety, efficiency, and structural performance across multiple engineering fields. Overall, the integration of advanced design principles, computational modeling, and modern manufacturing techniques enables mechanical metamaterials to provide high-performance, lightweight, and multifunctional structural solutions for next-generation engineering systems.

REVIEW QUESTIONS

1. What are mechanical metamaterials and how do they differ from conventional materials?
2. Define Poisson's ratio and explain the concept of negative Poisson's ratio.
3. What are auxetic materials and why are they important in engineering applications?
4. Explain the mechanism of re-entrant structures in auxetic metamaterials.
5. Describe the concept of elastic instability in metamaterial structures.
6. What is Euler buckling theory, and how does it relate to metamaterial design?
7. Explain the concept of snap-through instability.
8. What is the difference between bistable and multistable architectures?
9. Define negative stiffness and explain how it can be realized in metamaterials.
10. Why is negative stiffness useful in vibration isolation systems?
11. Derive the mathematical expression for Poisson's ratio.
12. Explain how unit-cell geometry affects the mechanical properties of metamaterials.
13. What factors influence the critical buckling load of a structure?
14. Explain the relationship between geometry and tunable stiffness in metamaterials.
15. How do hierarchical structures improve mechanical performance?
16. Discuss the role of auxetic metamaterials in protective helmet design.
17. Explain how mechanical metamaterials improve crashworthiness in automotive structures.

18 Describe the working principle of metamaterial-based aerospace panels.
19 How can metamaterials be used in earthquake-resistant structures?
20 Explain the application of shape morphing metamaterials in soft robotics.
21 Define specific energy absorption (SEA).
22 What is crushing force efficiency (CFE) and why is it important?
23 Explain how peak crushing force affects crashworthiness performance.
24 How is energy absorption capacity determined experimentally?
25 What factors influence the impact resistance of metamaterials?
26 Discuss the importance of programmable instability in metamaterial design.
27 Explain how computational modeling assists in metamaterial development.
28 Compare auxetic metamaterials with conventional cellular structures.
29 Discuss the advantages of hierarchical metamaterial architectures.
30 Explain the role of additive manufacturing in producing mechanical metamaterials.
31 Describe the working mechanism of a metamaterial-based automotive crash absorber.
32 Explain how auxetic helmet liners improve impact protection.
33 Discuss the advantages of metamaterial cores in aerospace sandwich panels.
34 How do seismic metamaterials reduce earthquake damage?
35 Explain how metamaterials enhance energy absorption in protective systems.
36 How can metamaterial structures be optimized for maximum energy absorption?
37 Discuss the potential of metamaterials in next-generation transportation systems.
38 Explain the role of topology optimization in mechanical metamaterial design.
39 How can smart materials enhance the functionality of metamaterials?
40 Discuss the challenges in large-scale manufacturing of metamaterials.
41 What are the emerging trends in mechanical metamaterial research?
42 How can machine learning assist in metamaterial design optimization?
43 Discuss the potential of multifunctional metamaterials in aerospace engineering.
44 Explain how programmable metamaterials can be used in adaptive structures.
45 Discuss future applications of metamaterials in biomedical engineering.
46 Calculate SEA if a structure absorbs 500 J of energy and weighs 2 kg.
47 Determine CFE if mean crushing force is 8 kN and peak force is 10 kN.
48 Explain how reducing peak crushing force improves crash safety.
49 Discuss how lattice geometry affects impact energy absorption.
50 Suggest design modifications to improve metamaterial crash absorbers.

4 Engineering Applications of Mechanical Metamaterials

4.1 INTRODUCTION

Mechanical metamaterials have attracted significant attention in recent years due to their ability to exhibit unconventional mechanical properties that arise from engineered architectures rather than intrinsic material composition. Through careful design of unit-cell geometry, topology, and hierarchical structures, these materials can achieve properties such as negative Poisson's ratio, tunable stiffness, enhanced energy absorption, and superior vibration damping. These characteristics make mechanical metamaterials highly suitable for advanced engineering systems where conventional materials fail to provide optimal performance (Surjadi et al., 2019).

The integration of mechanical metamaterials with modern manufacturing techniques such as additive manufacturing, microfabrication, and hybrid fabrication methods has enabled the practical realization of complex architected structures. As a result, metamaterials are increasingly being adopted in various engineering domains, including aerospace, automotive, civil engineering, electronics, and infrastructure development. Their ability to combine lightweight structural efficiency with multifunctional performance allows engineers to design next-generation materials capable of improving safety, durability, and operational efficiency.

This chapter focuses on the practical engineering applications of mechanical metamaterials, highlighting how their unique properties can be utilized to address real-world challenges. Key application areas include structural vibration control, impact-resistant aerospace and automotive components, flexible electronic systems, thermal and acoustic insulation structures, and load-bearing infrastructure elements.

4.2 STRUCTURAL VIBRATION CONTROL

Structural vibration is a common phenomenon encountered in engineering systems, such as buildings, bridges, aircraft structures, mechanical equipment, and transportation systems. Vibrations may arise due to external excitations such as wind loads, earthquakes, mechanical operations, or aerodynamic forces. Excessive vibrations can lead to structural fatigue, noise generation, reduced performance, and even structural failure. Therefore, effective vibration control is essential for improving the reliability and longevity of engineering structures.

Mechanical metamaterials provide innovative solutions for vibration control due to their engineered internal architectures and tunable mechanical properties. Unlike conventional damping materials that rely solely on intrinsic material properties,

DOI: 10.1201/9781003783671-4

metamaterials can manipulate the propagation of mechanical waves through periodic structures and resonant unit cells. These structures can create frequency bandgaps, which are specific frequency ranges where vibrational waves cannot propagate through the material. As a result, metamaterials can effectively block or attenuate unwanted vibrations.

One of the most important mechanisms used in vibration control is the concept of locally resonant metamaterials. These structures contain resonating elements that interact with incoming vibration waves and convert their energy into localized oscillations. This mechanism allows the structure to absorb or dissipate vibrational energy efficiently. Another approach involves negative stiffness metamaterials, which combine positive and negative stiffness elements to achieve enhanced vibration isolation and damping performance.

Mechanical metamaterials are increasingly used in seismic protection systems, vibration isolation platforms, aerospace structures, precision instruments, and mechanical equipment. For example, metamaterial-based vibration isolators can reduce ground vibrations transmitted to sensitive instruments, while metamaterial dampers can minimize oscillations in large structural systems such as bridges and tall buildings.

4.2.1 Working Principle

The working principle of structural vibration control using mechanical metamaterials is based on the manipulation and attenuation of mechanical wave propagation through engineered periodic structures. When a structure experiences external excitation such as wind loads, mechanical operations, or seismic disturbances, vibration waves propagate through the structural material. Mechanical metamaterials are designed with periodic unit cells or locally resonant structures that interact with these waves and modify their transmission characteristics.

When vibrational waves enter the metamaterial layer, the periodic architecture of the unit cells creates interference effects that disrupt the propagation of certain vibration frequencies. These structures can generate frequency bandgaps, which are specific frequency ranges in which vibration waves cannot propagate through the material. As a result, vibrations within these frequency ranges are significantly attenuated or completely blocked.

Another important mechanism involves locally resonant elements embedded within the metamaterial structure. These resonating components absorb incoming vibrational energy by oscillating at specific frequencies. The absorbed energy is then dissipated through internal damping mechanisms such as material friction and structural deformation. This process reduces the amplitude of vibrations transmitted to the surrounding structure.

In addition, some metamaterial systems incorporate negative stiffness or high-damping mechanisms, which enhance vibration isolation by reducing the effective stiffness of the system while maintaining stability. This allows the structure to isolate itself from external disturbances and maintain operational stability.

Table 4.1 presents major engineering applications of structural vibration control using mechanical metamaterials. The table highlights how different metamaterial

TABLE 4.1
Applications of Structural Vibration Control Using Mechanical Metamaterials

Application Area	Engineering System	Metamaterial Mechanism Used	Description of Application	Example
Seismic protection	Buildings and Bridges	Frequency bandgaps and damping layers	Metamaterial foundations or barriers reduce the transmission of seismic waves, protecting structures from earthquake-induced vibrations.	Metamaterial-based seismic isolators are used in earthquake-resistant building foundations.
Aerospace structures	Aircraft wings and fuselage	Locally resonant unit cells	Metamaterial panels absorb aerodynamic vibrations and reduce structural fatigue during flight.	Vibration-damping metamaterial panels in aircraft fuselage structures.
Precision machinery	Manufacturing equipment	Negative stiffness and vibration isolation	Metamaterial isolators prevent vibrations from reaching sensitive instruments, improving measurement accuracy.	Metamaterial vibration isolation platforms for semiconductor fabrication machines.
Railway systems	High-speed train tracks	Periodic lattice structures	Metamaterial layers placed under railway tracks attenuate vibration waves generated by train movement.	Vibration control layers beneath high-speed rail tracks.
Automotive systems	Vehicle chassis and engine mounts	Energy dissipation and resonant damping	Metamaterial dampers reduce vibrations from engines and road conditions, improving passenger comfort.	Metamaterial engine mounts and vibration-damping chassis components.
Industrial structures	Large rotating machinery	Resonant vibration absorbers	Metamaterial absorbers reduce mechanical oscillations in rotating equipment, increasing operational stability.	Metamaterial vibration dampers in turbine systems.

mechanisms, such as frequency bandgaps, local resonance, and enhanced damping, are utilized to control and attenuate vibrations in various engineering systems. These applications demonstrate the ability of metamaterials to improve structural stability, operational efficiency, and safety across fields such as civil engineering, aerospace, transportation, and industrial machinery.

4.3 IMPACT-RESISTANT AEROSPACE AND AUTOMOTIVE STRUCTURES

Impact resistance is a critical requirement in both aerospace and automotive engineering, where structures must withstand high-energy collisions, debris impacts, bird strikes, and crash events. Conventional materials such as solid metals or composite laminates can provide strength but often suffer from localized failure, high weight, and limited energy absorption capacity. Mechanical metamaterials offer an advanced alternative by utilizing engineered cellular architectures that enable controlled deformation and efficient energy dissipation.

Mechanical metamaterials used in impact-resistant structures typically consist of lattice frameworks, honeycomb structures, or hierarchical cellular geometries. These architectures distribute impact forces across a network of interconnected structural members, preventing stress concentration and allowing progressive collapse mechanisms such as bending, buckling, and plastic deformation. This controlled collapse enables the structure to absorb large amounts of kinetic energy while reducing the peak force transmitted to surrounding components.

In aerospace applications, metamaterial cores are commonly integrated into sandwich panels used in aircraft fuselage structures, satellite shielding systems, and spacecraft protective panels. These metamaterial cores provide excellent strength-to-weight ratios, allowing engineers to design lightweight structures that maintain high structural integrity under impact conditions. For example, lattice-core sandwich panels can protect aircraft structures from bird strikes or debris impacts during flight.

Similarly, in automotive engineering, mechanical metamaterials are increasingly used in vehicle crash absorbers, bumper systems, and crumple zones. The metamaterial architecture allows these components to deform progressively during collisions, absorbing crash energy and reducing the forces transmitted to passengers. Compared with conventional solid structures, metamaterial-based crash absorbers can significantly improve crashworthiness performance while reducing vehicle weight.

Another advantage of metamaterial-based impact-resistant structures is their ability to achieve high specific energy absorption (SEA), which is the energy absorbed per unit mass of the structure. By optimizing unit-cell geometry and material distribution, engineers can design metamaterials that absorb more impact energy than traditional materials without increasing structural mass.

Overall, mechanical metamaterials provide a powerful platform for developing lightweight, high-performance impact-resistant structures in aerospace and automotive systems. Their ability to combine energy absorption, structural efficiency, and tunable mechanical behavior makes them highly suitable for next-generation transportation and protective engineering applications.

4.3.1 Working Principle

The working principle of impact-resistant structures using mechanical metamaterials is based on the ability of engineered cellular architectures to absorb, redistribute, and dissipate kinetic energy during impact events. When an external object or collision force strikes a structure, a large amount of kinetic energy and stress waves are generated at the point of contact. In conventional solid materials, these forces often lead to localized stress concentration and sudden structural failure. In contrast, mechanical metamaterials are designed with periodic lattice or cellular structures that allow the impact energy to be distributed throughout the architecture.

When an impact occurs, the incoming kinetic energy first propagates through the lattice network as stress waves. The internal structural elements, such as beams, nodes, and ligaments, then begin to deform through mechanisms including bending, buckling, and rotation of unit cells. These deformation processes absorb a significant portion of the impact energy. As the impact force increases, the structure undergoes progressive collapse, where different layers or unit cells deform sequentially rather than failing simultaneously.

This progressive deformation converts the kinetic energy of the impact into internal strain energy and heat, thereby reducing the magnitude of the force transmitted to the protected structure. In many metamaterial designs, hierarchical or graded architectures further enhance this process by allowing different regions of the structure to deform at different stages of impact.

In aerospace and automotive applications, these mechanisms enable structures such as crash absorbers, sandwich panels, and protective shells to maintain structural integrity while minimizing damage. The ability to control deformation pathways ensures that impact energy is dissipated efficiently while keeping the structure lightweight. As a result, metamaterial-based impact-resistant structures provide improved crashworthiness, enhanced energy absorption, and better passenger or component protection compared with conventional structural materials.

Table 4.2 presents the major applications of impact-resistant structures using mechanical metamaterials in aerospace and automotive systems. The table highlights how various metamaterial architectures, such as lattice structures, honeycomb cores, auxetic designs, and graded density materials, improve energy absorption, reduce peak impact forces, and enhance structural safety in transportation and defense engineering.

4.4 FLEXIBLE ELECTRONICS AND WEARABLE SYSTEMS

Flexible electronics and wearable systems represent a rapidly growing field in modern technology, where electronic devices must operate reliably while undergoing bending, stretching, twisting, and compression. Conventional rigid electronic components often fail under large mechanical deformation, limiting their application in wearable devices. Mechanical metamaterials provide an effective solution by enabling structures with high flexibility, stretchability, and mechanical durability while maintaining structural integrity and functional performance.

TABLE 4.2
Applications of Impact-Resistant Aerospace and Automotive Structures Using Mechanical Metamaterials

Application Area	Engineering System	Metamaterial Structure Used	Working Mechanism	Example
Aircraft Structures	Aircraft fuselage and wings	Lattice-core sandwich panels	Cellular metamaterial cores absorb impact energy through progressive deformation and buckling of lattice members.	Octet-truss lattice cores used in aircraft fuselage panels for bird-strike protection.
Spacecraft Protection	Satellite shielding systems	Hierarchical lattice metamaterials	Multiple lattice layers dissipate energy from debris impacts in space environments.	Protective metamaterial shielding for satellites against micrometeoroid impacts.
Automotive Crash Systems	Vehicle bumpers and crumple zones	Energy-absorbing lattice structures	Metamaterial crash absorbers deform progressively during collisions, reducing peak force transmitted to passengers.	Metamaterial bumper systems in advanced electric vehicles.
Aerospace Sandwich Panels	Aircraft interior and structural panels	Honeycomb or lattice-core panels	Lightweight cellular cores distribute impact loads and prevent localized structural damage.	Aluminum honeycomb sandwich panels in aircraft flooring and cabin structures.
Protective Vehicle Components	Engine mounts and chassis components	Auxetic metamaterial structures	Auxetic structures expand laterally during compression, increasing energy absorption capability.	Auxetic lattice crash absorbers in automotive structural frames.
Military and Defense Vehicles	Armored vehicles and protective structures	Graded density metamaterials	Layered metamaterial structures absorb ballistic or blast energy through sequential collapse.	Lightweight metamaterial armor panels for military vehicles.

Mechanical metamaterials used in flexible electronics typically consist of periodic lattice structures, auxetic geometries, and serpentine interconnect architectures. These structures allow large mechanical deformation without causing excessive strain in the electronic components. By carefully designing the internal architecture, metamaterials can distribute mechanical stresses uniformly across the structure, preventing localized damage and enhancing the reliability of electronic devices.

Flexible metamaterial structures are commonly integrated with conductive materials, sensors, and electronic circuits to create wearable systems capable of monitoring physiological signals, environmental conditions, or human motion. For example, stretchable metamaterial substrates can support flexible circuits used in health monitoring devices, smart textiles, and wearable biomedical sensors. Their ability to conform to curved surfaces such as the human body significantly improves comfort and performance.

4.4.1 Working Principle

The working principle of flexible electronic systems based on mechanical metamaterials relies on the controlled deformation of architected structures that accommodate mechanical strain without damaging the electronic components. When external forces such as bending or stretching are applied, the metamaterial structure deforms through rotation, bending, and stretching of its unit cells rather than direct stretching of the electronic elements.

For instance, auxetic metamaterial structures expand laterally when stretched, allowing electronic components embedded within the structure to experience reduced localized strain. Similarly, serpentine or lattice-shaped conductive pathways can elongate through geometric deformation rather than material stretching. This mechanism allows the electronic circuit to maintain electrical functionality even when the substrate undergoes significant deformation.

Another important mechanism is strain distribution and energy absorption within the metamaterial architecture. The periodic geometry allows mechanical loads to be distributed across multiple structural elements, reducing stress concentration and preventing damage to sensitive electronic components.

As a result, metamaterial-based flexible electronic systems can maintain stable performance under repeated mechanical deformation, making them suitable for applications such as wearable health monitoring devices, flexible displays, smart clothing, electronic skin (e-skin), and biomedical implants.

Table 4.3 summarizes the major applications of flexible electronics and wearable systems enabled by mechanical metamaterials. The table highlights how different metamaterial architectures, such as auxetic structures, lattice substrates, and serpentine conductive networks, allow electronic devices to maintain functionality under large mechanical deformation. These technologies play a crucial role in health monitoring, smart textiles, flexible displays, and biomedical devices.

TABLE 4.3
Applications of Flexible Electronics and Wearable Systems Using Mechanical Metamaterials

Application Area	Wearable/ Electronic System	Metamaterial Structure Used	Working Mechanism	Example
Health Monitoring Devices	Wearable biosensors	Stretchable lattice or serpentine structures	Metamaterial substrate deforms through bending and rotation of unit cells, protecting electronic sensors from mechanical strain.	Wearable heart-rate and ECG monitoring patches.
Smart Textiles	Electronic fabrics	Auxetic metamaterial structures	Auxetic architecture expands laterally when stretched, allowing integrated electronic circuits to maintain electrical connectivity.	Smart clothing with embedded motion or temperature sensors.
Flexible Displays	Bendable electronic screens	Periodic lattice substrates	Metamaterial structures distribute mechanical stress during bending, preventing cracking of electronic layers.	Foldable smartphone displays and flexible LED screens.
Electronic Skin (E-Skin)	Artificial skin sensors	Micro-lattice metamaterials	Highly flexible architectures allow sensors to conform to curved surfaces while maintaining signal transmission.	Robotic skin sensors and prosthetic tactile sensing systems.
Wearable Motion Sensors	Body movement tracking devices	Stretchable serpentine interconnects	Conductive pathways deform geometrically rather than stretching directly, preserving electrical conductivity.	Fitness tracking bands and posture monitoring devices.
Biomedical Implants	Flexible medical electronics	Soft lattice metamaterial substrates	Metamaterial structures adapt to biological tissues, reducing mechanical mismatch and improving biocompatibility.	Flexible neural implants and implantable biosensors.

4.5 THERMAL AND ACOUSTIC INSULATION SYSTEMS

Thermal and acoustic insulation systems play an important role in engineering applications where it is necessary to control heat transfer and reduce unwanted sound transmission. Conventional insulation materials such as foams, fiberglass, and porous materials provide basic insulation properties, but their performance is often limited by material density, structural stability, and restricted control over wave propagation. Mechanical metamaterials offer a more advanced solution by using engineered architectures that manipulate thermal transport and acoustic wave propagation through structural design rather than relying solely on material composition (Fok et al., 2008).

In thermal insulation applications, metamaterials typically incorporate cellular or lattice architectures containing a large volume of air-filled voids. These voids significantly reduce the conduction of heat because air has low thermal conductivity. Additionally, the complex internal geometry of metamaterials increases the path length for heat transfer, thereby reducing the overall thermal flow through the structure. By carefully designing the geometry and material distribution, engineers can create metamaterials with low thermal conductivity and improved thermal insulation performance.

For acoustic insulation, metamaterials use periodic structures or resonant elements to control the propagation of sound waves. These structures can create acoustic bandgaps, which are specific frequency ranges where sound waves cannot propagate through the material. When sound waves encounter these metamaterial structures, they are either reflected, scattered, or absorbed by the internal architecture. This mechanism significantly reduces noise transmission and improves acoustic damping.

Mechanical metamaterials used for insulation often combine lightweight structures with multifunctional capabilities, allowing them to provide both structural support and insulation simultaneously. These materials are widely used in building construction, automotive cabins, aircraft interiors, industrial noise control systems, and thermal protection systems in aerospace engineering.

Overall, thermal and acoustic metamaterials provide significant advantages such as enhanced insulation performance, lightweight structural efficiency, tunable wave control, and multifunctional capabilities, making them highly suitable for modern engineering systems requiring advanced insulation technologies.

4.5.1 Working Principle

The working principle of thermal and acoustic insulation using mechanical metamaterials is based on controlling the transfer of heat and the propagation of sound waves through engineered structural geometries.

For thermal insulation, the metamaterial structure contains cellular networks and porous architectures that trap air within the structure. Since air has very low thermal conductivity, the heat transfer through the material is significantly reduced. Additionally, the complex pathways created by the metamaterial geometry increase the thermal resistance, slowing down heat flow across the structure.

For acoustic insulation, metamaterials utilize periodic structures and locally resonant elements to manipulate sound waves. When sound waves enter the metamaterial structure, the periodic architecture causes wave interference and scattering, which prevents certain frequencies of sound from propagating through the material. These frequency ranges are known as acoustic bandgaps.

In some metamaterial designs, resonant cavities or embedded oscillators absorb sound energy by vibrating at specific frequencies. The absorbed energy is then dissipated as heat through internal damping mechanisms.

Through these mechanisms, thermal resistance, acoustic bandgaps, wave scattering, and resonant absorption, mechanical metamaterials can provide highly efficient insulation systems for both thermal and acoustic environments.

Table 4.4 summarizes the major engineering applications of thermal and acoustic insulation systems using mechanical metamaterials. The table highlights how different metamaterial architectures, such as cellular lattices, honeycomb structures, and resonant acoustic metamaterials, provide improved insulation performance by controlling heat transfer and sound wave propagation. These systems are widely used in construction, automotive, aerospace, industrial machinery, and consumer electronics to enhance thermal efficiency and reduce noise.

TABLE 4.4
Applications of Thermal and Acoustic Insulation Systems Using Mechanical Metamaterials

Application Area	Engineering System	Metamaterial Structure Used	Working Mechanism	Example
Building Construction	Walls, ceilings, and floors	Cellular and porous lattice metamaterials	Air-filled voids reduce heat conduction, and periodic structures block sound transmission.	Acoustic metamaterial wall panels are used in soundproof rooms and auditoriums.
Automotive Industry	Vehicle cabins and engine compartments	Honeycomb or lattice insulation panels	Metamaterial structures absorb engine noise and reduce heat transfer from engine components.	Acoustic insulation panels in car dashboards and engine covers.
Aerospace Systems	Aircraft cabins and thermal protection systems	Lightweight lattice metamaterials	Cellular architectures minimize heat transfer and dampen noise from aerodynamic forces.	Thermal protection panels in aircraft interiors and spacecraft structures.

(Continued)

TABLE 4.4 (Continued)
Applications of Thermal and Acoustic Insulation Systems Using Mechanical Metamaterials

Application Area	Engineering System	Metamaterial Structure Used	Working Mechanism	Example
Industrial Noise Control	Machinery enclosures and factory walls	Resonant acoustic metamaterials	Locally resonant unit cells absorb mechanical vibration and reduce noise propagation.	Noise-reduction enclosures around turbines and compressors.
Consumer Electronics	Electronic device cooling and sound insulation	Micro-lattice metamaterial structures	Porous structures control heat dissipation and reduce acoustic vibrations in devices.	Acoustic insulation layers in laptops and high-performance electronics.
HVAC Systems	Air-conditioning ducts and ventilation systems	Periodic acoustic metamaterials	Acoustic bandgaps prevent noise transmission through ventilation channels.	Noise-reduction metamaterial ducts in building ventilation systems.

4.6 LOAD-BEARING INFRASTRUCTURE COMPONENTS

Load-bearing infrastructure components are structural elements designed to support and transmit loads safely within engineering structures such as buildings, bridges, towers, and transportation systems. Conventional construction materials like steel, concrete, and reinforced composites provide high strength but often result in heavy structures, high material consumption, and limited adaptability to changing loads or environmental conditions. Mechanical metamaterials offer a promising alternative by enabling lightweight structures with high strength, improved load distribution, and enhanced structural stability.

Mechanical metamaterials used in load-bearing applications typically employ cellular lattices, hierarchical architectures, and topology-optimized structures. These architectures allow the material to distribute loads across interconnected structural elements such as beams and nodes, rather than concentrating stresses in specific regions. This distributed load transfer mechanism improves structural efficiency and reduces the risk of local failure.

One of the major advantages of metamaterials in infrastructure systems is their high strength-to-weight ratio. By replacing solid materials with engineered lattice architectures, it is possible to significantly reduce structural weight while maintaining the required mechanical strength. This lightweight efficiency can lead to reduced material costs, easier transportation of structural components, and improved energy efficiency in construction projects.

Another important benefit is the ability of metamaterial structures to exhibit enhanced energy absorption and vibration damping. In infrastructure systems such as bridges or earthquake-resistant buildings, these properties can improve structural resilience by absorbing dynamic loads caused by wind, traffic, or seismic activity.

Mechanical metamaterials can also be designed with graded or hierarchical architectures, allowing different parts of the structure to possess different stiffness or strength levels depending on the expected load distribution. This adaptability improves structural reliability and performance under varying operational conditions.

Applications of metamaterial-based load-bearing components include lightweight bridge structures, advanced building frameworks, seismic-resistant foundations, structural panels, and transportation infrastructure systems. These materials have the potential to revolutionize civil engineering by enabling the development of stronger, lighter, and more resilient infrastructure systems.

4.6.1 Working Principle

The working principle of load-bearing metamaterial structures is based on efficient load distribution through engineered structural geometry. When an external load is applied to the structure, the internal lattice architecture transfers the load through a network of interconnected beams and nodes. Instead of concentrating stresses in a single region, the load is distributed across multiple structural paths.

This mechanism reduces stress concentration and improves the overall load-bearing capacity of the structure. Additionally, the cellular architecture allows the structure to deform slightly under load without causing structural failure, enhancing the stability and durability of the infrastructure component.

Through mechanisms such as bending, stretching, and controlled buckling of structural elements, metamaterial-based infrastructure components can maintain high structural performance while minimizing material usage. As a result, these structures achieve lightweight efficiency, improved mechanical stability, and enhanced structural resilience, making them suitable for modern infrastructure systems.

Table 4.5 presents major applications of load-bearing infrastructure components using mechanical metamaterials. The table highlights how metamaterial architectures such as lattice structures, hierarchical networks, and sandwich panels provide improved structural efficiency, load distribution, and resilience in modern civil engineering applications.

TABLE 4.5
Applications of Load-Bearing Infrastructure Components Using Mechanical Metamaterials

Application Area	Infrastructure System	Metamaterial Structure Used	Working Mechanism	Example
Bridge Structures	Bridge decks and supports	Lattice or cellular metamaterials	Distributed load transfer across lattice beams improves structural stability.	Lightweight metamaterial bridge panels.

(Continued)

TABLE 4.5 (Continued)
Applications of Load-Bearing Infrastructure Components Using Mechanical Metamaterials

Application Area	Infrastructure System	Metamaterial Structure Used	Working Mechanism	Example
Building Construction	Structural frameworks and beams	Hierarchical lattice structures	Hierarchical architectures distribute loads across multiple structural levels.	Metamaterial-based structural beams in high-rise buildings.
Seismic-Resistant Foundations	Building foundations	Energy-absorbing metamaterial layers	Cellular structures absorb and dissipate seismic vibrations.	Metamaterial seismic isolation foundations.
Transportation Infrastructure	Railway and highway supports	Topology-optimized lattice structures	Optimized architectures reduce weight while maintaining high strength.	Lightweight metamaterial rail support systems.
Structural Panels	Wall and floor systems	Honeycomb or lattice sandwich panels	Cellular cores provide high stiffness with low weight.	Metamaterial sandwich panels in modular construction.

4.7 CASE STUDY: SEISMIC METAMATERIALS FOR EARTHQUAKE PROTECTION IN URBAN INFRASTRUCTURE

4.7.1 Introduction

Earthquakes pose a significant threat to buildings and infrastructure, especially in seismic-prone regions. Traditional seismic protection techniques such as base isolation systems, dampers, and reinforced structural elements aim to reduce the damage caused by ground vibrations. However, these methods often rely on absorbing vibrations after they have already reached the structure. Mechanical metamaterials offer a novel approach by manipulating and attenuating seismic waves before they reach the building, providing an additional layer of protection.

4.7.2 Engineering Location Example

One of the most well-known demonstrations of seismic metamaterials was conducted near Grenoble, France, where researchers implemented a seismic metamaterial barrier consisting of a periodic array of cylindrical boreholes drilled into the ground. This experimental setup was designed to test how periodic structures could influence the propagation of seismic waves.

4.7.3 Design Concept

The metamaterial barrier consisted of several rows of cylindrical holes arranged periodically in the soil, forming a structure capable of interacting with seismic waves.

The periodic geometry created an artificial lattice within the ground, functioning similarly to phononic crystals that control wave propagation.

Key design parameters included the following:

- Periodic spacing of boreholes
- Controlled depth and diameter of the holes
- Arrangement forming a lattice-like seismic barrier

4.7.4 Working Mechanism

When seismic waves travel through the ground:

- The waves encounter the periodic metamaterial barrier.
- The periodic structure causes wave scattering and interference.
- Certain frequencies of seismic waves are reflected or attenuated due to the formation of seismic bandgaps.
- The amplitude of ground vibration reaching nearby structures is significantly reduced.

This mechanism effectively redirects or suppresses seismic energy, protecting buildings from strong ground motion.

4.7.5 Results and Observations

Field experiments showed that the metamaterial barrier significantly reduced surface wave amplitude and vibration intensity. Sensors placed behind the metamaterial structure recorded lower vibration levels compared with areas without the barrier.

4.8 CASE STUDY: AUXETIC METAMATERIALS IN SPORTS FOOTWEAR (NIKE FLYKNIT TECHNOLOGY)

4.8.1 Introduction

Sports footwear must provide comfort, flexibility, and impact protection while maintaining lightweight performance. Traditional shoe soles often struggle to balance flexibility and cushioning. Auxetic mechanical metamaterials have been introduced in modern footwear to enhance shock absorption and adaptability to foot movement.

4.8.2 Engineering Application

Companies such as Nike have incorporated auxetic-inspired lattice structures into advanced footwear designs like the Nike Flyknit and 3D-printed midsoles. These designs use cellular geometries that expand laterally under compression, improving energy distribution and cushioning.

4.8.3 Design Concept

The sole structure contains periodic auxetic unit cells fabricated using additive manufacturing techniques. The lattice geometry allows the sole to expand and contract in response to pressure applied during walking or running.

4.8.4 Working Mechanism

When a person steps on the shoe:

- The foot applies compressive force on the auxetic lattice.
- The lattice structure expands laterally due to its negative Poisson's ratio.
- Impact energy is distributed across the entire sole.
- The structure returns to its original shape after unloading.

4.8.5 Engineering Benefits

- Enhanced cushioning performance
- Improved load distribution across the foot
- Increased flexibility and comfort
- Lightweight structural design

4.8.6 Applications

- Sports footwear
- Athletic performance shoes
- Running shoes with energy-return midsoles

4.9 CASE STUDY: MECHANICAL METAMATERIALS IN AIRBUS AIRCRAFT STRUCTURES

4.9.1 Introduction

Aircraft structures require materials that provide high strength, lightweight performance, and impact resistance. Traditional solid materials increase aircraft weight, which negatively affects fuel efficiency. Mechanical metamaterials have emerged as promising candidates for aerospace structures due to their exceptional strength-to-weight ratios.

4.9.2 Engineering Application

Airbus and other aerospace manufacturers are investigating lattice-based metamaterial structures fabricated through additive manufacturing for aircraft components, such as interior panels, brackets, and energy-absorbing structures.

4.9.3 Design Concept

Aircraft panels are constructed using sandwich structures with metamaterial lattice cores placed between thin composite skins. The lattice core provides structural rigidity while minimizing weight.

4.9.4 Working Mechanism

During loading or impact:

- External forces are transferred to the lattice core.
- Structural members undergo bending and buckling deformation.
- Impact energy is dissipated through progressive structural collapse.
- The outer skins remain protected from damage.

4.9.5 Engineering Benefits

- Reduced aircraft weight
- Improved energy absorption capability
- Enhanced structural stiffness
- Increased fuel efficiency

4.9.6 Applications

- Aircraft fuselage panels
- Lightweight aerospace brackets
- Satellite structural components
- Spacecraft protective structures

4.10 CASE STUDY: ACOUSTIC METAMATERIAL NOISE BARRIERS FOR HIGHWAYS

4.10.1 Introduction

Traffic noise is a major environmental problem in urban areas, affecting the quality of life in residential zones located near highways. Conventional noise barriers often struggle to reduce low-frequency noise, which travels long distances.

4.10.2 Engineering Application

Acoustic metamaterial noise barriers have been developed to control sound waves using periodic resonant structures capable of creating acoustic bandgaps.

4.10.3 Design Concept

The noise barrier consists of periodic resonant cavities and lattice structures embedded within wall panels installed along highways.

4.10.4 Working Mechanism

When sound waves reach the barrier:

- Acoustic waves interact with resonant cavities in the metamaterial.
- The periodic structure creates acoustic bandgaps.
- Sound waves are reflected or absorbed within the structure.
- The transmitted noise level beyond the barrier is significantly reduced.

4.10.5 Engineering Benefits

- Effective reduction of low-frequency noise
- Lightweight structural panels
- Improved urban acoustic comfort
- Passive noise control without energy input

4.10.6 Applications

- Highway noise barriers
- Railway sound insulation systems
- Urban acoustic protection structures
- Industrial noise control

4.11 CASE STUDY: METAMATERIAL-BASED VIBRATION CONTROL IN METRO RAILWAY SYSTEMS (DELHI METRO)

4.11.1 Introduction

Urban metro railway systems generate significant ground vibrations due to high-speed train movement, wheel–rail interaction, and dynamic loading of tracks. These vibrations can propagate through the ground and affect nearby buildings, underground utilities, and sensitive equipment. Traditional vibration control methods, such as rubber pads, ballast mats, and floating slab tracks, provide partial mitigation but often have limitations in controlling a wide range of vibration frequencies. Mechanical metamaterials provide an innovative approach for controlling such vibrations through engineered periodic structures capable of manipulating wave propagation.

4.11.2 Engineering Location Example

The Delhi Metro Rail Corporation (DMRC) has implemented several advanced vibration mitigation technologies along metro lines that pass near sensitive zones such as hospitals, laboratories, and residential buildings. In research studies related to metro rail vibration mitigation, metamaterial-inspired periodic structures and wave barriers have been proposed as an effective method to reduce ground vibration transmission.

4.11.3 Design Concept

The metamaterial-based vibration control system consists of periodically arranged structural elements installed beneath or alongside railway tracks. These elements can take the form of:

- periodic concrete or steel columns
- buried cylindrical barriers
- lattice-based vibration isolation structures

The periodic arrangement creates a phononic crystal-like structure that interacts with vibration waves generated by train movement.

4.11.4 Working Mechanism

When metro trains move along the track:

- Dynamic forces generated at the wheel–rail contact create vibration waves in the ground.
- These waves propagate through the soil toward nearby buildings.
- The metamaterial barrier placed near the track interacts with the incoming vibration waves.
- The periodic structure produces wave scattering and vibration bandgaps.
- Vibrations within certain frequency ranges are attenuated or redirected away from sensitive structures.

As a result, the vibration amplitude reaching nearby infrastructure is significantly reduced.

4.11.5 Engineering Benefits

- Reduction of ground vibration levels near railway corridors
- Protection of nearby buildings and infrastructure
- Improved passenger comfort and environmental safety
- Passive vibration control without external energy input

4.12 SUMMARY

This chapter discussed the engineering applications of mechanical metamaterials, highlighting how their architected structures enable unique mechanical and multi-functional properties that cannot be achieved with conventional materials. The chapter began with structural vibration control, where metamaterial-based periodic structures and locally resonant systems can attenuate unwanted vibrations by creating frequency bandgaps and enhanced damping mechanisms. Such systems are particularly useful in buildings, transportation infrastructure, and precision engineering environments.

The chapter also examined the use of metamaterials in impact-resistant aerospace and automotive structures, where lattice architectures and cellular networks enable controlled deformation and efficient energy absorption during collisions or high-speed impacts. These structures improve crashworthiness while maintaining lightweight performance. Another important application discussed was flexible electronics and wearable systems, where metamaterial-based stretchable architectures allow electronic devices to maintain functionality even under large mechanical deformation, enabling applications in wearable health monitoring devices and flexible electronic systems.

Further, the chapter explored the use of metamaterials in thermal and acoustic insulation systems, where periodic and porous structures manipulate heat transfer and sound wave propagation. These materials can create acoustic bandgaps and reduce thermal conductivity, making them useful in building insulation, automotive noise control, and aerospace thermal protection systems. Finally, the chapter addressed the role of metamaterials in load-bearing infrastructure components, where engineered lattices and hierarchical structures provide high strength-to-weight ratios and improved load distribution, enabling lightweight yet durable structural systems.

Overall, the chapter demonstrates that mechanical metamaterials offer significant advantages in modern engineering applications by combining lightweight efficiency, multifunctional performance, enhanced energy absorption, and tunable mechanical behavior. These features make metamaterials highly promising for future technologies in aerospace, transportation, civil infrastructure, electronics, and advanced manufacturing.

REVIEW QUESTIONS

1. What are mechanical metamaterials, and why are they important for engineering applications?
2. Explain the role of metamaterials in structural vibration control.
3. What are vibration bandgaps and how do they help in vibration isolation?
4. Describe the advantages of metamaterials in aerospace structures.
5. How do metamaterials improve crashworthiness in automotive structures?
6. What are the key design features of flexible metamaterials used in wearable electronics?
7. Explain the importance of auxetic structures in wearable systems.
8. What are acoustic metamaterials, and how do they control sound waves?
9. Explain the concept of thermal insulation using metamaterial architectures.
10. What advantages do metamaterials offer for load-bearing infrastructure components?
11. Discuss the application of metamaterials in earthquake-resistant buildings.
12. Explain how metamaterials are used in aircraft fuselage structures.
13. Describe the role of metamaterials in automotive crash absorbers.
14. How can metamaterials improve vibration control in railway systems?
15. Explain the use of metamaterials in flexible electronic devices.
16. Discuss the application of acoustic metamaterials in highway noise barriers.

17 Explain the importance of metamaterials in smart textiles.
18 Describe how metamaterials can be used in thermal protection systems.
19 Discuss the role of metamaterials in lightweight bridge construction.
20 Explain how metamaterials contribute to sustainable infrastructure development.
21 Explain how periodic structures influence wave propagation in metamaterials.
22 Discuss the relationship between metamaterial architecture and mechanical performance.
23 How do lattice structures improve the strength-to-weight ratio of materials?
24 Explain the mechanism of energy absorption in metamaterial crash structures.
25 Discuss the role of hierarchical design in load-bearing metamaterials.
26 Compare conventional materials with metamaterials for vibration control applications.
27 Explain the advantages of metamaterials over traditional insulation materials.
28 Discuss the potential of metamaterials in next-generation aerospace structures.
29 Explain how additive manufacturing enables the fabrication of complex metamaterial architectures.
30 Discuss the role of metamaterials in sustainable engineering design.
31 Explain the working mechanism of seismic metamaterial barriers.
32 Discuss the design and benefits of metamaterial-based automotive crash absorbers.
33 Describe how auxetic metamaterials improve the performance of sports footwear.
34 Explain the use of metamaterials in aerospace sandwich panels.
35 Discuss the application of acoustic metamaterials in urban noise control.
36 What are the current challenges in the large-scale implementation of metamaterials?
37 How can computational design tools assist in optimizing metamaterial structures?
38 Discuss the future potential of programmable metamaterials in engineering systems.
39 Explain how machine learning can support metamaterial design optimization.
40 Discuss emerging trends in multifunctional metamaterials.
41 Explain how metamaterial barriers can reduce railway vibration transmission.
42 Discuss how auxetic materials improve shock absorption in wearable devices.
43 Suggest a metamaterial design suitable for highway noise reduction.
44 Explain how metamaterials can improve thermal insulation in aerospace systems.
45 Discuss the advantages of using lattice metamaterials in bridge structures.
46 How can metamaterials improve sustainability in infrastructure systems?
47 What limitations currently restrict the industrial adoption of metamaterials?
48 How can metamaterials contribute to smart city development?
49 Discuss the integration of metamaterials with smart materials for adaptive structures.
50 Suggest future engineering applications where metamaterials could play a major

5 Architectural Applications of Mechanical Metamaterials

5.1 INTRODUCTION

Architecture and building engineering continuously seek innovative materials and structural systems that can improve energy efficiency, structural performance, and environmental sustainability. Conventional construction materials, such as concrete, steel, glass, and composites, primarily rely on their intrinsic mechanical and thermal properties to achieve the desired building performance. However, modern architectural challenges, such as reducing energy consumption, improving acoustic comfort, and enhancing structural resilience, require materials with advanced multifunctional capabilities. Mechanical metamaterials provide a promising solution to these challenges by utilizing engineered architectures that control mechanical behavior, heat transfer, and wave propagation through a structural design.

Mechanical metamaterials derive their unique properties from periodic or architected microstructures, rather than from base materials alone. By carefully designing the geometry of these structures, it is possible to create materials with exceptional properties such as lightweight structural strength, enhanced thermal insulation, acoustic wave control, vibration isolation, and adaptive mechanical response. These features make metamaterials highly suitable for integration into architectural components, including walls, roofs, façades, structural frames, and interior systems.

In architectural engineering, mechanical metamaterials can significantly improve building performance in several ways. For example, cellular and lattice metamaterial structures can provide efficient thermal insulation, helping to regulate indoor temperature and reduce energy consumption for heating and cooling purposes. Similarly, acoustic metamaterials can be used to control sound propagation, improving indoor acoustic comfort in residential buildings, offices, and public spaces. In structural applications, lightweight metamaterial frameworks can be used to design large-span roofs, domes, and architectural frames that maintain high strength while reducing structural weight.

Another important advantage of metamaterials in architecture is their potential integration with smart and sustainable building systems. Metamaterial-based structures can be combined with sensors, adaptive materials, and energy-efficient design strategies to create buildings that respond dynamically to environmental conditions. Such systems contribute to the development of smart buildings and sustainable urban infrastructure.

DOI: 10.1201/9781003783671-5

Overall, the application of mechanical metamaterials in architecture represents an emerging interdisciplinary field that combines materials science, structural engineering, and architectural design. As fabrication technologies, like additive manufacturing, continue to advance, metamaterial-based architectural systems are expected to play a significant role in the development of energy-efficient, resilient, and aesthetically innovative buildings in the future (Siegmund & Barthelat, 2019).

5.2 THERMAL AND ACOUSTIC APPLICATIONS

Mechanical metamaterials have emerged as a powerful class of engineered materials capable of controlling thermal and acoustic energy in ways that conventional materials cannot. Their performance arises primarily from their architected microstructures rather than solely from their base material composition. By tailoring geometry, topology, and periodicity at different scales, mechanical metamaterials can manipulate heat flow, vibration, and sound propagation with high precision. These capabilities have opened new opportunities in architecture, aerospace, transportation, and electronic systems where thermal management and noise control are critical.

In thermal applications, metamaterials are designed to regulate heat conduction, convection, and radiation through structural engineering. Architected cellular structures, lattice networks, and layered configurations can significantly reduce the effective thermal conductivity, enabling improved thermal insulation and passive temperature regulation. Some metamaterials are also capable of creating thermal cloaking or directional heat flow, allowing heat to be guided around sensitive components without disturbing surrounding temperature fields. Such properties are particularly useful in buildings, energy systems, and electronic packaging, where maintaining stable temperatures is essential for performance and reliability.

Acoustic metamaterials, on the other hand, are specifically engineered to control the propagation of sound waves and mechanical vibrations. Unlike traditional sound-absorbing materials that rely mainly on bulk damping, acoustic metamaterials utilize periodic structures, resonant elements, and locally resonant units to manipulate sound at wavelengths much larger than the structural features. This allows them to exhibit unusual properties such as negative effective mass density, negative bulk modulus, and acoustic bandgaps, which prevent sound waves within certain frequency ranges from propagating through the material.

One of the most significant advantages of acoustic metamaterials is their ability to achieve high sound attenuation with lightweight and compact structures. This makes them highly suitable for applications such as noise barriers, vibration isolation systems, building acoustics, and transportation systems. For instance, metamaterial-based panels can be integrated into walls, ceilings, and partitions to reduce noise transmission while maintaining structural efficiency.

The combination of thermal and acoustic control within a single metamaterial structure is also gaining attention. Multifunctional metamaterials can simultaneously provide insulation and soundproofing, making them highly attractive for modern architectural and engineering systems. By integrating thermal management and acoustic attenuation capabilities into lightweight and adaptable structures, mechanical metamaterials offer a promising pathway toward more energy-efficient, comfortable, and sustainable built environments.

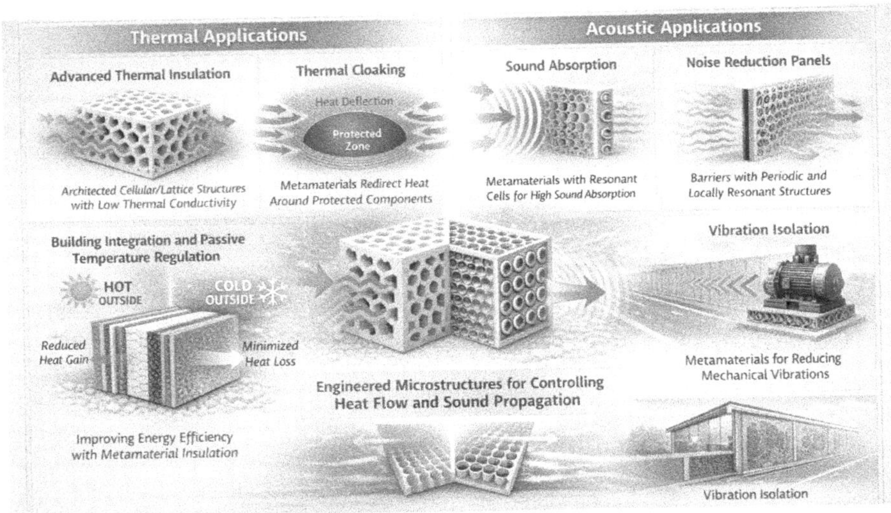

FIGURE 5.1 Thermal and acoustic applications of mechanical metamaterials.

Figure 5.1 illustrates the multifunctional capabilities of mechanical metamaterials in controlling both thermal energy and acoustic waves through engineered microstructures. The figure is divided into two main sections: thermal applications and acoustic applications, demonstrating how structural design enables advanced heat and sound management.

The thermal applications section highlights how metamaterials can regulate heat transfer through specially designed cellular and lattice structures. The first component shows advanced thermal insulation, where porous metamaterial architectures create tortuous pathways that reduce effective thermal conductivity. These structures limit heat flow by increasing thermal resistance and trapping air within internal cavities, which enhances insulation performance.

The next component illustrates thermal cloaking, where the metamaterial structure redirects heat flow around a protected region. Instead of heat passing directly through a sensitive area, the heat flux is guided around it, maintaining the temperature of the protected zone. This concept is particularly useful in applications involving electronic components, energy systems, and thermally sensitive structures.

Another part of the thermal section shows building integration and passive temperature regulation. Metamaterial-based insulation layers can be incorporated into building envelopes such as walls and roofs. These structures reduce heat gain during hot environmental conditions and minimize heat loss during colder conditions, thereby improving energy efficiency and indoor thermal comfort without requiring active energy consumption.

The acoustic applications section demonstrates how metamaterials can manipulate the propagation of sound waves and mechanical vibrations. The first example presents sound absorption, where resonant cavities within the metamaterial structure dissipate acoustic energy. These resonant units are capable of absorbing sound efficiently even at low frequencies, which are typically difficult to control using conventional materials.

The figure also depicts noise reduction panels, which use periodic and locally resonant structures to block or attenuate sound waves across specific frequency ranges. Such panels can be used in buildings, transportation systems, and industrial environments to reduce noise pollution and improve acoustic comfort.

Another application shown in the figure is vibration isolation, where metamaterials are used to suppress mechanical vibrations generated by machinery or infrastructure. Through carefully designed resonant elements and periodic structures, these materials can prevent vibration transmission to surrounding structures, thereby protecting sensitive equipment and improving structural stability.

At the center of the figure, the engineered metamaterial structure symbolizes the core concept of microstructural design controlling both heat flow and sound propagation. This demonstrates the multifunctional nature of mechanical metamaterials, where a single engineered architecture can simultaneously provide thermal management and acoustic control.

5.2.1 Thermal Insulation and Passive Temperature Control

Thermal insulation and passive temperature control are essential aspects of modern architectural and engineering design, particularly in the context of energy-efficient buildings and advanced structural systems. Mechanical metamaterials provide innovative solutions for thermal management due to their engineered microstructures and tunable thermal properties. Unlike conventional insulating materials that rely primarily on the low thermal conductivity of the base material, metamaterials utilize architectural design and geometric configurations to control the transfer of heat.

Thermal insulation refers to the reduction of heat transfer between objects or environments with different temperatures. Heat transfer generally occurs through conduction, convection, and radiation, and effective insulation materials are designed to minimize these modes of transfer. Mechanical metamaterials achieve this by incorporating porous cellular structures, lattice architectures, and hierarchical frameworks that significantly reduce the effective thermal conductivity of the material. The presence of internal voids and complex pathways increases the thermal resistance, thereby reducing heat flow.

Architected metamaterial structures, such as honeycomb lattices, auxetic frameworks, and periodic cellular materials, are particularly effective for thermal insulation. These structures create long and tortuous heat conduction paths, which reduce the rate of heat transfer. Additionally, the trapped air within the porous structures acts as a natural insulator, further enhancing the overall thermal performance. Because air has very low thermal conductivity, the combination of solid framework and air-filled cavities leads to highly efficient insulation systems (Phoenix & Wilson, 2018).

Passive temperature control is another important advantage offered by mechanical metamaterials. Passive systems regulate temperature without the need for external power sources or active mechanical systems. Through careful design of internal geometry, metamaterials can manage heat flow by directing, delaying, or redistributing thermal energy. For example, gradient lattice structures can guide heat away from sensitive areas, while layered metamaterials can act as thermal barriers that stabilize indoor temperatures.

In architectural applications, metamaterial-based thermal insulation can be integrated into walls, roofs, façades, and building envelopes. These materials help maintain indoor comfort by reducing heat gain during hot conditions and minimizing heat loss during cold conditions. As a result, buildings require less energy for heating and cooling, leading to improved energy efficiency and reduced environmental impact.

Another emerging concept is thermal cloaking, where metamaterials are designed to redirect heat around a protected region without disturbing the external temperature field. Such designs can protect sensitive components, improve thermal stability in electronics, and enhance the performance of thermal management systems in advanced engineering applications.

Figure 5.2 illustrates the concept of thermal insulation and passive temperature control using mechanical metamaterials and highlights how engineered microstructures can effectively regulate heat transfer in architectural and engineering systems.

The first section of the figure (Porous Metamaterial Structure) presents a cellular or lattice-based metamaterial block containing multiple internal voids. These porous structures create long and tortuous pathways for heat conduction, which significantly reduces the effective thermal conductivity of the material. The trapped air within the cavities further enhances insulation because air has very low thermal conductivity. As a result, the metamaterial structure slows down heat transfer from one side to the other.

The second section of the figure (thermal cloaking) demonstrates the concept of heat deflection around a protected zone. In this configuration, the metamaterial is designed to redirect heat flow around a specific region rather than allowing it to pass directly through it. This phenomenon is known as thermal cloaking, where the internal region remains thermally isolated while the external heat flow continues without

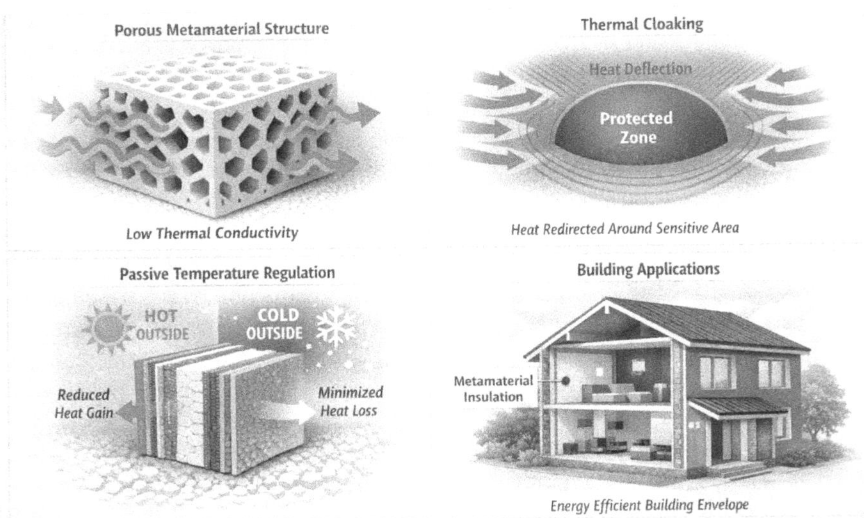

FIGURE 5.2 Thermal Insulation and passive temperature control using mechanical metamaterials.

major disturbance. Such designs are useful for protecting temperature-sensitive components in electronic devices, buildings, and energy systems.

The third section of the figure (passive temperature regulation) shows how metamaterial layers help control temperature differences between external and internal environments. When the outside environment is hot, the metamaterial layer reduces heat gain into the interior space. Conversely, during cold conditions, the structure minimizes heat loss, maintaining a more stable indoor temperature. This passive control occurs without requiring active energy input, making it an energy-efficient thermal management strategy.

The fourth section of the figure (Building Applications) illustrates the integration of metamaterial-based insulation within the building envelope, including walls and structural panels. By embedding these materials into architectural elements, buildings can achieve improved thermal performance, reduced energy consumption for heating and cooling, and enhanced indoor comfort.

5.2.2 Acoustic Metamaterials and Soundproofing

Acoustic metamaterials are a class of engineered materials specifically designed to control, manipulate, and attenuate sound waves and vibrations through carefully structured geometries. Unlike conventional soundproofing materials that rely primarily on bulk mass or porous absorption, acoustic metamaterials achieve superior performance by utilizing periodic architectures, resonant elements, and subwavelength structural features. These structural characteristics enable precise control over sound propagation, reflection, and absorption across specific frequency ranges.

The fundamental working principle of acoustic metamaterials is based on wave–structure interaction. When sound waves interact with specially designed microstructures, phenomena such as local resonance, bandgap formation, and negative effective material properties can occur. These properties allow metamaterials to block or significantly reduce sound transmission within selected frequency ranges known as acoustic bandgaps. As a result, sound waves that fall within these frequency bands are unable to propagate through the material.

One of the key mechanisms used in acoustic metamaterials is local resonance, where small resonant units or cavities are embedded within the material structure. These resonators vibrate at specific frequencies and absorb or dissipate incoming acoustic energy. This mechanism allows acoustic metamaterials to achieve effective low-frequency sound attenuation, which is typically difficult to obtain using conventional soundproofing materials such as foams or fibrous absorbers (Jang et al., 2022).

Acoustic metamaterials can be designed in several architectural forms, including membrane-type structures, Helmholtz resonator arrays, labyrinthine channels, and periodic lattice structures. Membrane-based metamaterials use thin elastic membranes with attached masses to create resonant behavior that absorbs sound energy. Helmholtz resonator structures consist of cavities connected by narrow necks that trap and dissipate sound waves at specific frequencies. Labyrinthine metamaterials use complex internal pathways to slow down and scatter sound waves, effectively reducing noise transmission.

These materials are widely used in soundproofing and noise control applications across various sectors. In architectural engineering, acoustic metamaterials are incorporated into walls, ceilings, partitions, and façade panels to reduce indoor noise levels and improve acoustic comfort in residential and commercial buildings. In transportation systems, such as automobiles, aircraft, and trains, metamaterial-based panels are used to reduce cabin noise and vibration.

Industrial environments also benefit from acoustic metamaterials through their use in noise barriers, machinery enclosures, and vibration-damping systems. These structures help protect workers from excessive noise exposure and improve the operational stability of equipment. Additionally, acoustic metamaterials are increasingly applied in auditoriums, recording studios, and performance halls to optimize sound quality and control unwanted reflections.

Another emerging area is the development of acoustic cloaking devices, where sound waves are redirected around an object without scattering, effectively making the object acoustically invisible. Although still largely in the research stage, this concept demonstrates the potential of metamaterials to revolutionize sound control technologies.

Figure 5.3 illustrates the working principles and applications of acoustic metamaterials for soundproofing and noise control. The figure highlights how engineered microstructures can manipulate sound waves through mechanisms such as local resonance, acoustic bandgaps, and periodic structural design, enabling efficient attenuation of sound and vibration.

The first section of the figure (local resonance) demonstrates the use of embedded resonant units within the metamaterial structure. These resonators are designed to oscillate at specific frequencies when exposed to incident sound waves. As a result, the resonant elements absorb and dissipate acoustic energy, preventing sound from propagating further through the material. This mechanism is particularly effective for attenuating low-frequency noise, which is typically difficult to control using traditional sound-absorbing materials.

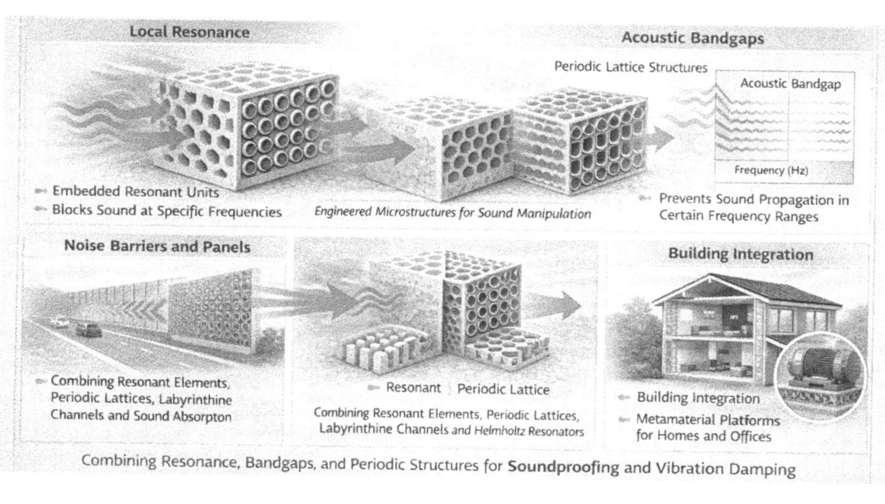

FIGURE 5.3 Acoustic metamaterials and soundproofing.

The second section of the figure (acoustic bandgaps) shows how periodic lattice structures can create frequency ranges in which sound waves cannot propagate. These ranges are known as acoustic bandgaps. When sound waves fall within these specific frequencies, they are either reflected or strongly attenuated by the metamaterial structure. The periodic arrangement of cavities or lattice elements disrupts the propagation of acoustic waves, providing highly efficient sound insulation.

The third section of the figure (Noise Barriers and Panels) illustrates the application of acoustic metamaterials in noise control systems such as roadside barriers and soundproof panels. These structures combine resonant cavities, periodic lattices, and labyrinthine channels to absorb and scatter sound waves. By integrating these features, metamaterial panels can significantly reduce noise levels generated by traffic, industrial machinery, or urban activities.

Another part of the figure shows engineered resonant lattice structures that combine multiple sound-attenuation mechanisms, such as resonance and wave scattering. These hybrid architectures enhance sound absorption across a wider frequency range, making them suitable for advanced acoustic engineering applications.

The final section of the figure (building integration) highlights how acoustic metamaterials can be incorporated into building structures such as walls, façades, and partitions. These materials help reduce noise transmission between indoor and outdoor environments, improving acoustic comfort in residential, commercial, and industrial spaces. Additionally, metamaterial structures can be used for vibration isolation in machinery and equipment, preventing unwanted mechanical vibrations from spreading through structural components.

5.2.3 Integration into Walls, Roofs, and Interior Architectural Elements

Mechanical metamaterials are increasingly being integrated into architectural components such as walls, roofs, ceilings, and interior partitions to enhance both thermal and acoustic performance. Unlike conventional construction materials, metamaterials rely on engineered geometries and periodic microstructures to control the transfer of heat and sound within buildings. Their lightweight nature, structural adaptability, and multifunctional capabilities make them highly suitable for modern architectural design focused on energy efficiency, comfort, and sustainability.

One of the primary areas of integration is within building walls and façades. Metamaterial-based panels can be incorporated into multilayer wall systems to provide improved thermal insulation and sound attenuation. Cellular lattice structures or porous metamaterials embedded within wall panels create tortuous pathways that reduce heat conduction and absorb acoustic energy. This helps maintain stable indoor temperatures while simultaneously minimizing the transmission of external noise into interior spaces.

Roof systems also benefit significantly from metamaterial integration. Roofs are major pathways for solar heat gain and environmental noise, particularly in urban environments. By incorporating metamaterial layers into roofing structures, it is possible to enhance thermal resistance and acoustic damping. Architected metamaterial layers can reduce the penetration of heat during hot weather while preventing heat loss during cooler periods. At the same time, these materials can attenuate noise generated by rainfall, wind, or external environmental sources.

Interior architectural elements, such as ceilings, partitions, acoustic panels, and decorative wall systems, can also incorporate metamaterial structures. Acoustic metamaterial panels installed in ceilings or interior walls can effectively control sound reflections and reverberations, improving the acoustic quality of spaces such as conference rooms, classrooms, auditoriums, and offices. These structures are often designed with resonant cavities, periodic channels, or labyrinthine pathways that trap and dissipate sound energy.

Another advantage of metamaterial-based architectural components is their design flexibility and aesthetic adaptability. Because the performance of metamaterials is largely determined by their geometry rather than solely by material composition, designers can tailor shapes and patterns to achieve both functional and visual architectural goals. This allows metamaterials to be integrated into decorative façades, patterned interior panels, and modular building elements without compromising structural or aesthetic requirements.

Additionally, metamaterials can support the development of multifunctional building envelopes that simultaneously provide structural support, insulation, and acoustic control. Such integrated systems reduce the need for multiple separate materials, resulting in lighter construction, simplified installation, and improved overall building efficiency.

Figure 5.4 illustrates the practical integration of mechanical metamaterials into various architectural components, highlighting their role in improving thermal insulation, acoustic control, and overall building energy efficiency. The figure presents four key architectural applications, namely enhanced wall insulation, acoustic metamaterial ceilings and partitions, advanced roofing systems, and multifunctional building envelopes.

The upper-left section (enhanced wall insulation) shows a multilayer wall structure incorporating metamaterial panels. These panels contain engineered porous or

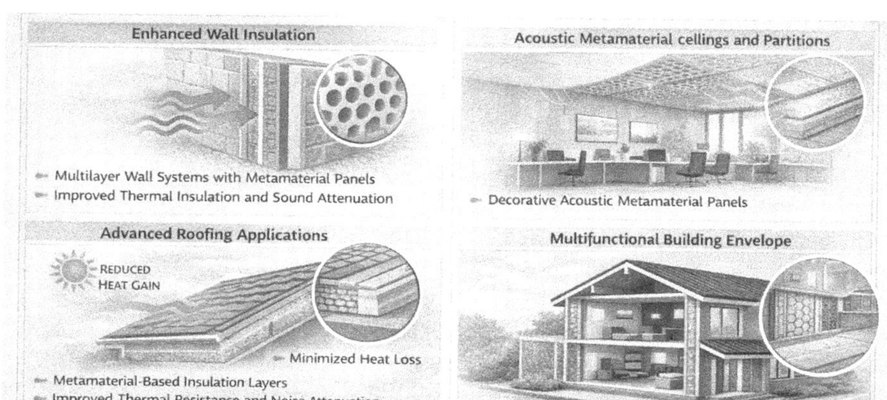

FIGURE 5.4 Integration of mechanical metamaterials into walls, roofs, and interior architectural elements.

lattice microstructures that create complex pathways for heat transfer. As a result, heat conduction through the wall is significantly reduced. Additionally, the cellular architecture helps absorb and attenuate sound waves, providing both thermal insulation and noise reduction within buildings.

The upper-right section (acoustic metamaterial ceilings and partitions) illustrates the use of metamaterial-based panels in interior architectural elements such as ceilings and partitions. These panels typically contain periodic cavities or resonant microstructures that trap and dissipate sound energy. Such systems are particularly useful in offices, conference rooms, classrooms, and public buildings where sound control and acoustic comfort are important. Decorative metamaterial panels also offer architectural flexibility while maintaining functional performance.

The lower-left section (Advanced Roofing Applications) highlights the role of metamaterials in roof insulation systems. Roofs are major pathways for heat gain during hot weather and heat loss during cold conditions. By integrating metamaterial-based insulation layers within roofing structures, the transfer of thermal energy is significantly reduced. The figure indicates reduced heat gain from solar radiation and minimized heat loss, thereby contributing to improved indoor temperature stability and energy efficiency.

The lower-right section (Multifunctional Building Envelope) presents a complete building structure, where metamaterials are incorporated into the external envelope of the building. In this configuration, metamaterial layers simultaneously provide thermal insulation, acoustic attenuation, and structural functionality. The building envelope acts as an integrated system that controls heat flow, reduces external noise, and enhances overall indoor environmental quality.

5.3 STRUCTURAL APPLICATIONS IN ARCHITECTURE

Mechanical metamaterials have gained significant attention in architectural engineering due to their unique mechanical properties derived from engineered microstructures rather than material composition alone. These architected materials enable innovative structural solutions by providing enhanced strength-to-weight ratios, deformation control, energy absorption, and adaptive mechanical behavior. As a result, mechanical metamaterials are increasingly being explored for use in architectural components such as load-bearing elements, adaptive façades, lightweight structural panels, and impact-resistant systems.

One of the key advantages of structural metamaterials is their ability to achieve high stiffness and strength while maintaining low density. Architected lattice structures, such as honeycomb, truss-based lattices, and periodic cellular frameworks, distribute loads efficiently across the structure. This load distribution capability allows structures to maintain mechanical stability while significantly reducing overall weight. In architectural construction, such lightweight yet strong structures are valuable for reducing material usage and improving structural efficiency.

Auxetic metamaterials, which exhibit a negative Poisson's ratio, are particularly useful in architectural structural systems. When subjected to tensile forces, these materials expand laterally instead of contracting. This unusual deformation behavior enhances indentation resistance, shear stiffness, and energy absorption capacity.

In building structures, auxetic metamaterials can improve resistance against mechanical stresses, making them suitable for applications such as protective panels, structural reinforcements, and vibration-resistant architectural elements.

Mechanical metamaterials can also be designed to provide impact resistance and energy absorption, which is important for structures exposed to dynamic loads such as earthquakes, strong winds, or accidental impacts. Cellular metamaterial structures deform in controlled ways under external loads, dissipating mechanical energy and preventing catastrophic structural failure. Such properties make them useful for shock-absorbing structural components and protective building systems.

Another emerging application is the use of metamaterials in adaptive and responsive architectural structures. Architected materials can be designed to change shape or stiffness in response to environmental conditions such as temperature, mechanical stress, or humidity. These properties allow buildings to incorporate adaptive façades and morphing structural elements that respond to changing environmental conditions while maintaining structural stability.

Structural metamaterials also support the development of modular and prefabricated construction systems. Because their mechanical performance is largely governed by geometry, these materials can be manufactured using advanced fabrication methods such as additive manufacturing, laser cutting, and modular assembly techniques. This enables the creation of complex structural panels and components that can be easily integrated into modern building systems.

5.3.1 Lightweight Domes and Roof Structures

Lightweight domes and roof structures are important architectural elements used to cover large spaces such as auditoriums, stadiums, exhibition halls, transportation terminals, and industrial facilities. Traditionally, these structures rely on materials such as steel trusses, reinforced concrete shells, or space frames. However, the integration of mechanical metamaterials has opened new possibilities for creating lightweight, strong, and structurally efficient roofing systems.

Mechanical metamaterials are particularly suitable for dome and roof structures because their mechanical behavior is governed by engineered geometry and internal architecture rather than material composition alone. Architected lattice structures, cellular frameworks, and periodic truss networks can distribute loads efficiently while significantly reducing the overall weight of the structure. This enables the construction of large-span roofs with minimal material usage while maintaining structural stability.

One of the key advantages of metamaterial-based roofing systems is their high strength-to-weight ratio. Lightweight lattice architectures, such as octet trusses, honeycomb structures, and hierarchical cellular frameworks, provide excellent load-bearing capabilities while reducing structural mass. This reduction in weight lowers the demand on supporting columns and foundations, which can result in more economical and efficient construction.

Metamaterial structures also improve the load distribution and deformation behavior of dome-shaped structures. Domes inherently distribute loads along curved surfaces, and when combined with metamaterial lattice frameworks, the load transfer becomes even more efficient. The interconnected network of struts and nodes allows forces to be distributed across the entire structure, reducing stress concentration and improving overall structural resilience.

Another important benefit is the energy absorption capability of metamaterial lattices. Cellular metamaterials can deform in controlled ways under external loads such as wind pressure, seismic forces, or impact events. This property helps prevent sudden structural failure by dissipating energy through localized deformation within the lattice structure.

In addition to structural benefits, metamaterial roof systems can provide multifunctional performance, including thermal insulation, acoustic damping, and ventilation. Porous lattice structures allow air circulation and reduce heat accumulation under large roof surfaces. When combined with acoustic metamaterial panels, these systems can also minimize sound reflections and reverberation in large indoor spaces.

Advanced fabrication technologies such as additive manufacturing, modular prefabrication, and digital construction methods make it possible to produce complex metamaterial lattice components for dome and roof systems. Modular metamaterial panels can be assembled into large structural frameworks, allowing faster construction and improved design flexibility.

Figure 5.5 illustrates the application of mechanical metamaterials in the design of lightweight dome and roof structures for architectural systems. The figure demonstrates how architected lattice frameworks can be used to construct large-span roofing systems that are structurally efficient, lightweight, and multifunctional.

The left section of the figure presents a large dome structure supported by a metamaterial lattice framework. This lattice consists of interconnected struts arranged in a periodic geometric pattern, forming a highly efficient load-bearing network. Such structures distribute applied loads evenly throughout the framework, reducing localized stress

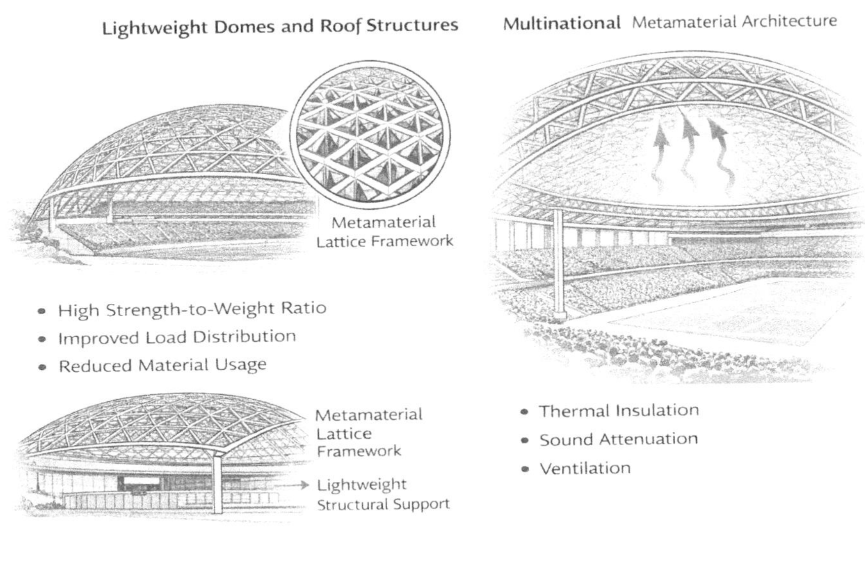

FIGURE 5.5 Lightweight domes and roof structures using mechanical metamaterials.

concentrations. As a result, the dome can maintain high structural stability while using significantly less material compared to traditional solid or heavy truss-based roofing systems. The illustration highlights the key advantages of metamaterial-based roof structures, including a high strength-to-weight ratio, improved load distribution, and reduced material usage.

The figure also shows a magnified view of the metamaterial lattice architecture, emphasizing the engineered microstructure responsible for the mechanical performance of the structure. These periodic cellular frameworks act as structural skeletons that efficiently transfer loads along multiple pathways. This structural configuration enables large-span roof systems that are both lightweight and mechanically robust.

The lower portion of the left section further demonstrates how the metamaterial lattice framework provides lightweight structural support for buildings such as stadiums, exhibition halls, or large public spaces. Because of the reduced structural weight, the loads transmitted to supporting columns and foundations are minimized, leading to more economical and efficient architectural designs.

The right section of the figure highlights the multifunctional capabilities of metamaterial-based roof systems. In addition to structural support, the architected lattice structures can also provide additional functionalities such as thermal insulation, sound attenuation, and natural ventilation. The porous nature of the metamaterial architecture allows controlled airflow, which can help regulate temperature inside large enclosed spaces. At the same time, the internal cavities within the lattice structures help absorb and dissipate sound energy, improving acoustic conditions in large indoor environments.

5.3.2 Earthquake-Resilient Structural Designs

Earthquake-resilient structural design is a critical aspect of modern architecture, particularly in regions that are prone to seismic activity. Traditional earthquake-resistant structures rely on strategies such as base isolation, energy-dissipating dampers, and flexible structural frames to reduce damage during seismic events. Recently, mechanical metamaterials have emerged as a promising approach for enhancing the seismic performance of buildings due to their engineered microstructures and controllable mechanical behavior.

Mechanical metamaterials can be designed to exhibit tailored stiffness, controlled deformation, and high energy absorption capacity, which are essential characteristics for seismic protection. During an earthquake, structures experience dynamic loads and ground vibrations that generate stress waves traveling through the building. Metamaterial-based structural components can absorb, redirect, or dissipate these vibrational energies, thereby reducing the transmission of destructive forces to the main structural framework.

One of the most promising metamaterial concepts for seismic protection involves periodic lattice structures and seismic metamaterials embedded within building foundations or structural elements. These engineered structures can create elastic wave bandgaps, which prevent seismic waves within certain frequency ranges from propagating through the structure. By blocking or attenuating these frequencies, the metamaterial layer acts as a protective barrier that shields the building from strong ground vibrations.

Auxetic metamaterials are particularly useful for earthquake-resilient structural systems. Due to their negative Poisson's ratio, auxetic structures expand laterally when stretched and contract uniformly when compressed. This behavior increases shear resistance, indentation resistance, and structural stability, allowing buildings to better withstand multidirectional forces generated during seismic activity (Yan et al., 2026).

Another important mechanism is energy dissipation through controlled deformation. Cellular metamaterial structures can deform elastically under seismic loads and recover their original shape once the load is removed. This reversible deformation helps absorb large amounts of mechanical energy without causing permanent damage to the structural system. As a result, metamaterial-based structural components can act as shock absorbers within the building framework.

Metamaterials can also be incorporated into foundation systems, seismic barriers, and structural joints. For example, metamaterial-based seismic barriers installed around a building can redirect incoming seismic waves away from the structure. Similarly, lattice-based structural components integrated into beams and columns can improve the overall ductility and vibration-damping capacity of the building.

In addition to structural safety, metamaterial-based earthquake-resilient systems contribute to lightweight construction and improved structural efficiency. Their geometry-driven mechanical behavior allows engineers to design structures that are both strong and flexible, enabling buildings to tolerate significant ground motion without catastrophic failure.

Figure 5.6 illustrates the concept of earthquake-resilient structural systems using mechanical metamaterials, demonstrating how engineered lattice architectures can mitigate the destructive effects of seismic waves. The figure is divided into two main sections: seismic protection mechanisms and earthquake-resilient building implementation.

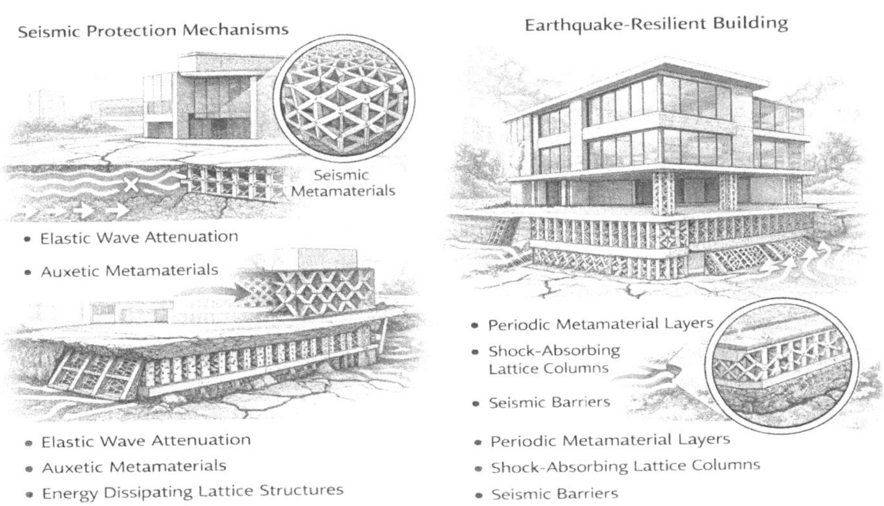

FIGURE 5.6 Earthquake-resilient structural designs using mechanical metamaterials.

The left section of the figure highlights the fundamental seismic protection mechanisms provided by metamaterials. The illustration shows seismic waves propagating through the ground and interacting with a layer of seismic metamaterials embedded in the foundation or surrounding soil. These periodic metamaterial structures are designed to manipulate elastic wave propagation. One of the primary mechanisms shown is elastic wave attenuation, where the metamaterial layer reduces the intensity of seismic waves before they reach the building structure.

Another important concept presented is the use of auxetic metamaterials, which exhibit a negative Poisson's ratio. When subjected to stress, auxetic structures expand laterally rather than contracting, increasing their ability to absorb energy and resist deformation. This property enhances structural stability during multidirectional seismic loading.

The figure also illustrates energy-dissipating lattice structures, which deform in a controlled manner when exposed to dynamic loads. These lattice frameworks act as structural dampers, absorbing and dissipating seismic energy to prevent excessive vibrations from propagating through the building.

The right section of the figure shows a conceptual earthquake-resilient building that incorporates metamaterial-based structural elements. The building foundation includes periodic metamaterial layers designed to reduce the transmission of seismic waves from the ground to the superstructure. These layers function similarly to seismic shields by redirecting or attenuating incoming ground vibrations.

In addition, the building columns incorporate shock-absorbing lattice structures, which act as energy-absorbing components capable of reducing structural vibrations during seismic events. The figure also depicts seismic barriers surrounding the foundation, which are metamaterial-based underground structures designed to deflect or dissipate incoming seismic waves before they reach the building.

5.3.3 Lightweight Architectural Frames and Lattice Structures

Lightweight architectural frames and lattice structures are increasingly used in modern construction to achieve efficient load-bearing performance while minimizing structural weight. Mechanical metamaterials provide innovative opportunities in this area because their mechanical behavior is governed primarily by engineered geometric architectures rather than only material composition. By designing periodic lattice frameworks and cellular structures, architects and engineers can create structural systems that are both lightweight and mechanically robust (Wu et al., 2023).

Architectural lattice structures consist of interconnected struts and nodes arranged in periodic or hierarchical patterns, forming an open cellular framework. These structures distribute loads efficiently throughout the network, allowing forces to be transferred along multiple pathways. As a result, lattice frameworks can support significant loads while using much less material compared to conventional solid structures. This high strength-to-weight ratio makes them particularly attractive for large-span architectural applications such as roofs, bridges, pavilions, and façade systems.

Mechanical metamaterials enable the design of advanced lattice architectures such as octet trusses, tetrahedral frameworks, and hierarchical cellular lattices. These structures exhibit exceptional mechanical properties, including high stiffness, enhanced load distribution, and improved resistance to buckling or deformation.

Because the internal architecture governs mechanical performance, engineers can tune the structural response by adjusting parameters such as cell geometry, lattice density, and strut orientation.

Another advantage of metamaterial-based lattice structures is their ability to provide multifunctional performance. In addition to structural support, open cellular frameworks can contribute to thermal management, acoustic damping, and natural ventilation within architectural spaces. The porous nature of lattice structures allows air circulation, which helps regulate temperature in enclosed environments. At the same time, internal cavities can help dissipate sound energy, improving acoustic comfort.

Lightweight architectural frames are also beneficial for modular and prefabricated construction systems. Lattice-based metamaterial panels can be fabricated using advanced manufacturing techniques such as additive manufacturing, laser cutting, and modular assembly methods. These prefabricated components can then be assembled into large structural frameworks, reducing construction time and improving design flexibility.

Furthermore, the geometric flexibility of lattice metamaterials allows architects to create aesthetically appealing and structurally efficient designs. Complex geometric patterns can be incorporated into building façades, interior structures, and roof systems without compromising structural integrity. This integration of structural performance with architectural aesthetics enables innovative designs that combine functionality, efficiency, and visual appeal.

5.3.4 Soundproofing and Vibration Isolation in Buildings

Soundproofing and vibration isolation are essential considerations in modern architectural design, particularly in urban environments where buildings are exposed to traffic noise, industrial vibrations, and mechanical equipment disturbances. Mechanical metamaterials offer innovative solutions for controlling sound and structural vibrations due to their engineered microstructures and wave-manipulation capabilities. Unlike conventional soundproofing materials that rely mainly on mass and damping, metamaterials can manipulate acoustic waves and mechanical vibrations through periodic structural design (Zhao et al., 2024).

Acoustic metamaterials used in building systems often incorporate periodic resonant cavities, lattice frameworks, or membrane-based structures that interact with incoming sound waves. These engineered architectures can create acoustic bandgaps, which are specific frequency ranges where sound propagation is strongly suppressed. When sound waves encounter such structures, they are either absorbed, reflected, or scattered, resulting in a significant reduction in noise transmission through walls, floors, and ceilings.

One important mechanism for soundproofing is local resonance. In metamaterial-based acoustic panels, small resonant elements vibrate at particular frequencies when exposed to incident sound waves. This vibration absorbs acoustic energy and converts it into mechanical energy within the structure, effectively reducing the sound intensity transmitted through the building element. Such systems are particularly effective for controlling low-frequency noise, which is typically difficult to attenuate using traditional sound insulation materials.

In addition to soundproofing, mechanical metamaterials play a crucial role in vibration isolation within buildings. Vibrations generated by mechanical systems such as heating, ventilation, and air conditioning (HVAC) equipment, elevators, generators, or nearby transportation systems can propagate through structural elements and cause discomfort or structural fatigue. Metamaterial-based vibration isolators can be designed to attenuate mechanical waves and reduce vibration transmission through structural components.

Periodic lattice structures embedded within building foundations or structural supports can act as vibration-damping systems. These architected materials distribute vibrational energy across multiple structural pathways, where it is gradually dissipated through internal deformation. In some cases, metamaterial structures can also redirect vibration waves away from sensitive areas of a building.

Mechanical metamaterials can be integrated into various architectural components, including walls, floors, ceilings, façades, and structural supports. Acoustic metamaterial panels installed in ceilings and interior partitions help reduce reverberation and improve sound quality in spaces such as offices, lecture halls, recording studios, hospitals, and residential buildings. Similarly, metamaterial-based floor systems can isolate vibrations between different levels of a building.

Another advantage of metamaterial-based soundproofing systems is their lightweight and compact design. Traditional sound insulation methods often require thick layers of dense materials, whereas metamaterials achieve similar or better performance using geometrically engineered structures with lower material usage. This makes them highly suitable for modern architectural designs where space efficiency and structural weight are important considerations.

5.3.5 Acoustic Noise Control in Urban Buildings

Urban buildings are frequently exposed to high levels of environmental noise generated by traffic systems, construction activities, industrial operations, railways, and aircraft. Continuous exposure to such noise can negatively affect human health, productivity, and overall indoor comfort. As urban populations continue to grow and cities become denser, effective strategies for acoustic noise control in buildings are becoming increasingly important. Mechanical metamaterials provide innovative solutions for mitigating urban noise through their engineered structures capable of manipulating sound waves.

Traditional noise control methods typically rely on heavy insulation materials, thick walls, and damping layers to reduce sound transmission. While effective to some extent, these approaches often require large amounts of material and space. In contrast, acoustic metamaterials utilize periodic architectures, resonant cavities, and labyrinthine channels to control sound propagation more efficiently. These engineered structures can create acoustic bandgaps, which prevent sound waves within certain frequency ranges from passing through the material.

One of the most effective metamaterial-based solutions for urban noise control involves the use of acoustic façade panels. These panels can be integrated into building exteriors to reduce the transmission of external noise into indoor spaces. The internal microstructures within these panels interact with incoming sound waves by absorbing, reflecting, or scattering acoustic energy, thereby significantly lowering noise levels inside the building.

Metamaterial-based noise barriers and façade systems are particularly effective for buildings located near busy roads, railways, or airports. By carefully designing the geometry of the metamaterial structure, engineers can target specific noise frequencies commonly produced by traffic or industrial machinery. This selective attenuation allows the building envelope to act as an acoustic shield against urban noise pollution.

Another promising application is the incorporation of acoustic metamaterials in windows and ventilation systems. Traditional building ventilation openings often allow external noise to enter indoor environments. Metamaterial-based acoustic vents and window inserts can reduce noise transmission while still allowing air circulation. These structures often employ labyrinthine pathways or resonant cavities that slow down and dissipate sound waves while maintaining airflow.

In addition to façade systems, metamaterials can be integrated into interior acoustic panels, ceilings, and partition walls to further enhance indoor acoustic conditions. Such installations are particularly beneficial in residential apartments, office buildings, hospitals, schools, and hotels, where quiet indoor environments are essential for comfort and productivity.

Mechanical metamaterials also provide the advantage of lightweight and compact noise control systems. Because their performance is governed by structural geometry rather than material thickness alone, they can achieve high acoustic attenuation with significantly less material. This makes them well suited for modern architectural designs that require space efficiency and aesthetic flexibility.

5.3.6 Vibration Isolation Systems for Building Foundations

Vibration isolation is an important aspect of structural design in modern buildings, particularly in urban environments where structures are frequently exposed to ground vibrations generated by traffic, railways, industrial machinery, and construction activities. Excessive vibrations can affect structural stability, occupant comfort, and the performance of sensitive equipment. Mechanical metamaterials provide advanced solutions for vibration isolation by utilizing engineered periodic structures that manipulate and attenuate mechanical waves before they propagate through building foundations.

Traditional vibration isolation systems typically rely on rubber pads, spring-based isolators, and damping materials placed between the building foundation and the ground. While these methods are effective for certain vibration frequencies, they often have limitations in controlling broadband or low-frequency vibrations. Mechanical metamaterials, however, offer enhanced vibration control through periodic lattice architectures and resonant structural elements that can be tuned to target specific vibration frequencies.

One of the primary mechanisms used in metamaterial-based vibration isolation systems is the formation of elastic wave bandgaps. Periodic metamaterial structures embedded within the foundation or surrounding soil can block the propagation of mechanical waves within certain frequency ranges. When ground vibrations encounter these engineered layers, the waves are either reflected, scattered, or attenuated, thereby preventing them from reaching the building structure.

Another important approach involves the use of lattice-based metamaterial columns or foundation layers. These cellular structures can deform elastically when subjected to dynamic loads, absorbing and dissipating vibrational energy. The internal geometry of the lattice framework distributes vibration forces across multiple pathways, reducing the intensity of vibrations transmitted to the building superstructure.

Metamaterial-based seismic barriers can also be installed around building foundations to redirect or dissipate incoming ground vibrations. These barriers consist of periodic arrays of engineered structures embedded in the soil surrounding the building. By altering the propagation path of seismic or vibrational waves, these metamaterial barriers help protect the building from excessive dynamic loading.

In addition to vibration control, metamaterial-based foundation systems offer lightweight structural support and improved mechanical resilience. Because their mechanical properties are governed by structural geometry, engineers can tailor the lattice architecture to achieve optimal combinations of stiffness, flexibility, and energy dissipation.

These vibration isolation systems are particularly beneficial for buildings that house precision equipment, research laboratories, hospitals, and high-tech manufacturing facilities, where even small vibrations can interfere with operations. They are also valuable for buildings located near subway systems, highways, or industrial zones where continuous vibration exposure is common.

5.3.7 Metamaterial-Based Floor and Ceiling Systems

Metamaterial-based floor and ceiling systems represent an emerging architectural solution that integrates structural support, acoustic control, and vibration isolation within building interiors. In modern buildings, floors and ceilings are not only required to support loads but also to manage sound transmission, mechanical vibrations, and thermal comfort between different levels of a structure. Mechanical metamaterials provide advanced capabilities in these areas due to their engineered lattice architectures and tunable mechanical properties.

Traditional floor and ceiling systems typically consist of concrete slabs, steel beams, wooden frameworks, and acoustic insulation layers. While these systems provide structural stability, they often require additional materials such as heavy insulation layers or damping materials to control sound and vibrations. Metamaterial-based systems offer a more integrated approach by using architected cellular structures that simultaneously perform structural and functional roles.

One of the major advantages of metamaterial floor systems is their ability to provide efficient load distribution with reduced structural weight. Lattice frameworks composed of interconnected struts and nodes create a lightweight yet strong structure capable of supporting loads across large floor spans. These periodic architectures distribute loads along multiple pathways, improving structural stability and resistance to deformation while minimizing material usage.

In addition to structural support, metamaterial floors can significantly reduce vibration transmission between building levels. Vibrations generated by walking, machinery, or external sources can travel through conventional floor structures and cause discomfort or noise disturbances. Metamaterial lattice layers embedded within floor systems can absorb and dissipate these vibrations through controlled deformation and energy dissipation mechanisms.

Ceiling systems incorporating acoustic metamaterials can effectively control sound reflections, reverberations, and airborne noise transmission. Periodic cavities and resonant elements within metamaterial ceiling panels interact with sound waves to absorb acoustic energy. This makes them particularly suitable for offices, classrooms, auditoriums, conference halls, and residential buildings, where maintaining a comfortable acoustic environment is essential.

Another important feature of metamaterial-based floor and ceiling systems is their ability to create acoustic bandgaps. These bandgaps prevent sound waves within specific frequency ranges from propagating through the structure, thereby reducing noise transmission between different floors. This capability is especially valuable in multi-story buildings, hotels, hospitals, and apartment complexes, where sound insulation between floors is critical.

Metamaterial-based systems also support modular and prefabricated construction methods. Floor and ceiling panels with integrated lattice architectures can be manufactured using advanced fabrication techniques such as additive manufacturing, modular assembly, and advanced composite fabrication. These prefabricated elements can then be installed quickly during construction, reducing building time and improving structural efficiency.

Figure 5.7 illustrates the integration of metamaterial-based floor and ceiling systems within a multi-level building structure, highlighting their role in acoustic control, vibration isolation, and lightweight structural support. The figure presents a

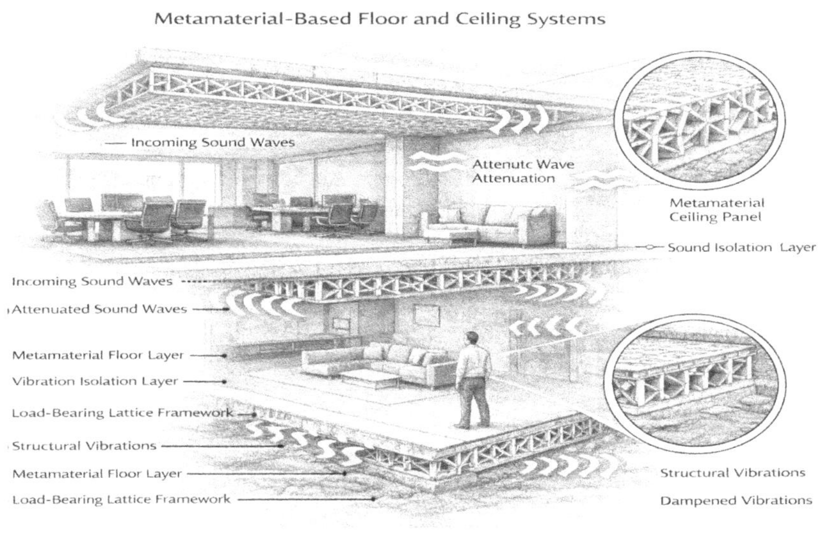

FIGURE 5.7 Metamaterial-based floor and ceiling systems in buildings.

cross-sectional view of a building interior, demonstrating how engineered metamaterial layers can be incorporated into architectural elements to enhance building performance.

The upper section of the figure shows a metamaterial ceiling panel installed in the ceiling of an interior space. This panel consists of a periodic lattice architecture designed to interact with incoming sound waves. When sound waves generated within the building travel toward the ceiling, the metamaterial structure attenuates the acoustic energy through wave scattering, local resonance, and internal damping mechanisms. As indicated by the arrows in the figure, the incoming sound waves are partially absorbed and converted into weaker, attenuated waves, thereby reducing noise levels within the room and improving indoor acoustic comfort.

The sound isolation layer incorporated within the ceiling system further enhances the acoustic performance by preventing sound transmission between different floors. This combination of metamaterial lattice structures and acoustic layers helps control both reverberation within a room and noise propagation between building levels.

The lower section of the figure illustrates a metamaterial-based floor system that supports the structural loads of the upper level while simultaneously providing vibration isolation. The floor structure includes a metamaterial floor layer composed of a lattice framework that distributes loads efficiently across the structure. This load-bearing lattice framework provides structural stability while maintaining a lightweight design.

The figure also indicates the presence of a vibration isolation layer within the floor structure. Vibrations generated by human movement, machinery, or external sources can propagate through conventional floors and cause disturbances in lower levels of a building. The metamaterial lattice architecture absorbs and dissipates these vibrations through controlled deformation of the structural elements, thereby reducing the amplitude of vibrations transmitted through the building.

5.4 INTEGRATION WITH SMART AND SUSTAINABLE BUILDING SYSTEMS

The rapid development of smart technologies and sustainable construction practices has significantly transformed modern architectural design. Buildings are no longer viewed merely as static structures but as dynamic systems capable of adapting to environmental conditions, optimizing energy consumption, and improving occupant comfort. Mechanical metamaterials offer promising opportunities for integration with smart and sustainable building systems due to their tunable structural properties, multifunctional capabilities, and adaptability to advanced sensing and control technologies.

One of the key advantages of mechanical metamaterials in sustainable architecture is their ability to provide passive environmental control. Architected materials with cellular or lattice microstructures can regulate heat transfer, airflow, and acoustic propagation without requiring active energy input. When incorporated into building envelopes such as walls, roofs, and façades, metamaterials can significantly reduce heating and cooling demands, thereby improving energy efficiency and reducing carbon emissions.

Metamaterials can also be combined with smart sensors and monitoring systems to create adaptive building structures. Sensors embedded within metamaterial components can monitor parameters such as temperature, vibration, structural stress, and acoustic levels. This real-time data allows building management systems to adjust environmental controls, structural responses, or maintenance strategies accordingly. For example, smart metamaterial panels could adapt their stiffness or configuration in response to structural loads or environmental changes.

Another important aspect of smart building integration is the use of adaptive or responsive metamaterials. Certain architected materials can change their mechanical or thermal properties when exposed to external stimuli such as temperature, pressure, or electromagnetic fields. These responsive metamaterials can enable self-adjusting building components, such as façades that regulate airflow, shading systems that respond to solar radiation, or structures that adjust stiffness to counteract dynamic loads.

In sustainable building systems, metamaterials can also support renewable energy integration and resource efficiency. Lightweight lattice structures reduce material consumption while maintaining structural strength, contributing to more sustainable construction practices. Additionally, metamaterial-based components can improve the performance of solar panels, ventilation systems, and thermal energy management systems integrated into building envelopes.

Metamaterials are also well suited for modular and prefabricated construction approaches, which are increasingly used in sustainable architecture. Prefabricated metamaterial panels can be manufactured with precise geometric control and assembled quickly on-site, reducing construction waste and improving overall efficiency. Such systems are particularly beneficial for green buildings designed to meet sustainability certifications and energy efficiency standards.

Furthermore, the multifunctional nature of metamaterials allows a single architectural element to perform multiple roles, such as structural support, thermal insulation, soundproofing, and vibration control. This integration reduces the need for separate materials and layers, resulting in lighter structures and improved material efficiency.

5.4.1 Energy-Efficient Building Envelopes

The building envelope plays a critical role in determining the overall energy performance, thermal comfort, and environmental impact of modern buildings. It acts as the interface between the indoor environment and external climatic conditions, controlling the transfer of heat, air, moisture, and sound. With increasing emphasis on sustainable construction and energy conservation, the development of energy-efficient building envelopes has become a key objective in architectural engineering. Mechanical metamaterials offer innovative solutions for improving the performance of building envelopes through engineered structural architectures that regulate energy flow.

Traditional building envelopes typically consist of multilayered systems incorporating insulation materials, structural panels, vapor barriers, and façade elements. While these systems provide basic thermal protection, they often require thick insulation layers and additional energy-consuming systems to maintain comfortable indoor conditions. Mechanical metamaterials, however, can achieve improved thermal control by utilizing periodic cellular structures and lattice architectures that reduce effective thermal conductivity.

Metamaterial-based building envelopes often incorporate porous or cellular microstructures that create tortuous pathways for heat conduction. These structures slow down heat transfer through the building walls and roofs, helping maintain stable indoor temperatures during both hot and cold conditions. As a result, buildings require less energy for HVAC, leading to improved energy efficiency and reduced operational costs.

Another important feature of metamaterial envelopes is their ability to support passive temperature regulation. Architected materials can be designed to control the direction and rate of heat flow through the envelope. For instance, some metamaterial structures can reflect or redirect solar radiation during hot conditions while retaining internal heat during colder periods. This adaptive thermal behavior contributes to maintaining indoor comfort without relying heavily on mechanical climate control systems.

Metamaterial-based building envelopes can also integrate acoustic insulation and vibration-damping properties. Urban buildings are frequently exposed to external noise sources such as traffic, construction, and industrial activities. Cellular metamaterial layers embedded within façade panels can absorb and attenuate sound waves, providing improved acoustic comfort in indoor environments.

In addition to thermal and acoustic functions, metamaterial envelopes offer advantages in terms of lightweight construction and structural efficiency. Lattice-based panels reduce the overall weight of façade systems while maintaining adequate mechanical strength. This reduces the load on supporting structural components and allows for more flexible architectural designs.

Mechanical metamaterials are also compatible with advanced smart building technologies. When combined with sensors and responsive systems, metamaterial envelopes can adapt their properties based on environmental conditions. For example, smart metamaterial façades may adjust airflow, shading, or thermal conductivity in response to temperature variations or solar intensity.

Figure 5.8 illustrates the concept of energy-efficient building envelopes incorporating mechanical metamaterials to enhance thermal insulation, acoustic control, and structural efficiency. The diagram presents a cross-sectional view of a multi-story building where metamaterial lattice structures are integrated into the external façade and structural layers of the building envelope.

The upper section of the figure shows solar radiation from the external environment interacting with the building façade. The metamaterial-based envelope contains engineered lattice microstructures that regulate the transfer of heat through the building walls. These architected structures create tortuous pathways for heat conduction, which reduce thermal transfer from the external environment to the interior space. As a result, the indoor environment remains thermally stable even under strong solar exposure, reducing the energy demand for air-conditioning systems.

The figure also highlights thermal insulation layers within the metamaterial envelope. These layers consist of cellular or porous metamaterial architectures that trap air and increase thermal resistance. By slowing down heat flow across the façade, the building envelope acts as an effective barrier against external temperature fluctuations.

FIGURE 5.8 Energy-efficient metamaterial-based building envelopes.

Another important feature illustrated in the diagram is acoustic control. Urban buildings are often exposed to environmental noise such as traffic and industrial activities. The internal cavities within the metamaterial lattice structures interact with incoming sound waves, scattering and attenuating acoustic energy. This process reduces noise transmission into indoor spaces, thereby improving acoustic comfort for building occupants.

The figure also demonstrates the use of metamaterial-based façade panels integrated into the structural framework of the building. These panels provide multiple functions simultaneously, including structural support, thermal insulation, and sound attenuation. Because metamaterial structures rely on engineered geometry rather than bulk material mass, they can achieve high mechanical performance while maintaining a lightweight structural design.

In addition, the lightweight lattice framework illustrated along the building envelope reduces the overall weight of the façade system. This reduction in structural weight decreases the load on supporting structural elements such as beams and columns, enabling more efficient building design and improved material utilization.

5.4.2 Smart Façades and Adaptive Architectural Systems

Smart façades and adaptive architectural systems represent an important advancement in modern building design, enabling structures to respond dynamically to environmental conditions and optimize building performance. Unlike conventional static façades, smart façade systems are capable of adjusting their thermal, optical, and structural properties in response to changes in temperature, solar radiation, wind, and occupancy conditions. Mechanical metamaterials provide an innovative platform for developing such adaptive building components due to their geometry-driven mechanical behavior and tunable physical properties.

Mechanical metamaterials can be engineered to form adaptive façade structures that regulate heat transfer, airflow, and solar radiation. Architected lattice structures embedded within façade panels can modify the way energy interacts with the building envelope. For example, metamaterial-based façade elements can be designed to expand, contract, or reconfigure their geometry in response to external stimuli such as temperature changes or mechanical loads. This capability allows buildings to regulate heat gain and improve indoor comfort without relying heavily on energy-intensive climate control systems.

One important function of smart façades is solar radiation management. During periods of intense sunlight, metamaterial façade structures can act as adaptive shading systems, reducing solar heat gain and glare inside buildings. Conversely, during colder conditions or low sunlight periods, the façade can allow greater solar penetration to assist with passive heating. Such adaptive behavior helps improve building energy efficiency and thermal comfort.

Metamaterial-based smart façades can also enhance natural ventilation and airflow control. Cellular or porous lattice structures can guide airflow through the building envelope, allowing controlled ventilation while maintaining thermal insulation. This feature helps regulate indoor temperature and reduce reliance on mechanical ventilation systems.

In addition to thermal management, adaptive façade systems can provide acoustic noise reduction in urban environments. Metamaterial panels integrated into building exteriors can absorb and attenuate sound waves generated by traffic, construction activities, or industrial sources. By controlling acoustic wave propagation, these façade systems improve the indoor acoustic environment of buildings located in densely populated cities.

Another advantage of metamaterial-based adaptive systems is their compatibility with smart sensing and building automation technologies. Sensors embedded within façade panels can monitor environmental parameters such as temperature, humidity, solar intensity, and wind speed. This information can be used by building management systems to adjust façade configurations automatically, ensuring optimal building performance under varying environmental conditions.

Furthermore, metamaterial smart façades support lightweight construction and architectural flexibility. Their lattice-based structures reduce material usage while maintaining structural strength, enabling innovative façade designs with complex geometries and aesthetic appeal.

5.4.3 Sustainable Construction Using Metamaterials

Sustainable construction has become a fundamental objective in modern architectural engineering due to increasing concerns about energy consumption, environmental impact, and resource depletion. Buildings account for a significant portion of global energy use and carbon emissions, which has driven the need for innovative materials and design strategies that improve efficiency while minimizing environmental harm. Mechanical metamaterials offer promising opportunities for sustainable construction because their performance is governed by engineered structural geometry rather than solely by material composition, enabling optimized use of materials and enhanced multifunctional capabilities.

One of the primary contributions of metamaterials to sustainable construction is their ability to enable lightweight structural systems. Architected lattice structures, cellular frameworks, and periodic geometries can achieve high strength and stiffness with significantly reduced material usage. This reduction in material consumption lowers the environmental footprint associated with material extraction, manufacturing, and transportation. Lightweight metamaterial structures also decrease the load on building foundations and supporting components, leading to more efficient overall construction.

Metamaterials also contribute to sustainability through improved thermal insulation and energy management. Cellular and porous metamaterial structures can reduce heat transfer across building envelopes by creating complex conduction pathways and trapping insulating air pockets. When integrated into walls, roofs, and façade systems, these materials help maintain stable indoor temperatures, reducing the need for energy-intensive heating and cooling systems.

Another important aspect of sustainable construction is multifunctionality. Traditional building systems often require multiple layers of materials to achieve structural strength, insulation, soundproofing, and vibration control. Metamaterials, however, can combine several of these functions within a single engineered structure. For example, a metamaterial panel may simultaneously provide structural support, thermal insulation, acoustic attenuation, and vibration damping. This multifunctional capability reduces the need for additional materials, resulting in more efficient and environmentally friendly construction systems.

Metamaterials are also compatible with advanced manufacturing and modular construction techniques. Technologies such as additive manufacturing, digital fabrication, and prefabricated modular assembly enable the precise production of complex metamaterial architectures. These methods reduce construction waste, improve manufacturing efficiency, and allow buildings to be assembled more quickly with fewer resources.

In addition, metamaterial-based systems support the integration of renewable energy technologies and passive environmental control strategies. Architected structures can enhance airflow, optimize solar exposure, and improve heat distribution within buildings. These capabilities contribute to the development of energy-efficient and climate-responsive architectural systems.

Metamaterials also enable the design of recyclable and adaptable construction components. Because their mechanical performance depends on geometry rather than bulk material properties alone, metamaterial structures can be produced using sustainable or recyclable materials such as advanced composites, polymers, or lightweight metals. Furthermore, modular metamaterial components can be easily replaced or reconfigured, extending the lifespan of building structures and reducing long-term environmental impact.

5.5 CASE STUDIES OF METAMATERIALS IN ARCHITECTURAL ENGINEERING

The application of mechanical metamaterials in architectural engineering has progressed from theoretical research to practical implementation in several experimental and real-world projects. These case studies demonstrate how engineered material architectures can enhance structural performance, thermal efficiency, acoustic control, and sustainability in modern buildings. By examining these examples, architects and engineers can better understand the potential of metamaterials to transform architectural design and construction.

One of the earliest architectural explorations of metamaterials involved the use of auxetic structures in façade panels. Auxetic metamaterials, characterized by their negative Poisson's ratio, expand laterally when stretched and contract uniformly when compressed. Researchers have developed façade systems incorporating auxetic lattice patterns that allow buildings to adapt to mechanical stresses, wind loads, and temperature changes. These adaptive façade panels can improve structural resilience while also creating visually distinctive architectural patterns.

Another notable case study involves the use of acoustic metamaterials in building interiors. Several research laboratories and experimental architecture projects have implemented acoustic metamaterial panels in auditoriums, lecture halls, and office environments. These panels use periodic resonant cavities and labyrinthine channels to absorb and attenuate sound waves across targeted frequency ranges. As a result, they significantly improve indoor acoustic performance while maintaining lightweight and aesthetically pleasing designs.

Metamaterial-based thermal insulation systems have also been explored in experimental building envelopes. In these systems, porous or cellular metamaterial structures are integrated into wall panels to reduce heat transfer through the building façade. By creating tortuous pathways for heat conduction, these structures enhance thermal resistance and help maintain stable indoor temperatures. Such designs demonstrate how metamaterials can contribute to energy-efficient building envelopes and passive temperature control.

Another emerging example is the development of seismic metamaterials for earthquake protection. Researchers have proposed and experimentally tested periodic metamaterial structures embedded in building foundations or surrounding soil. These structures can create seismic wave bandgaps, which prevent certain frequencies of seismic waves from propagating toward the building. Although still largely at the experimental stage, such systems show significant promise for improving earthquake resilience in architectural structures.

Metamaterials have also been applied in the design of lightweight architectural frameworks and large-span structures. Lattice-based metamaterial frameworks have been used in experimental pavilions, exhibition structures, and research prototypes. These architected frameworks achieve high structural strength while minimizing material usage, demonstrating the potential of metamaterials for sustainable and resource-efficient construction.

Additionally, research in metamaterial-based vibration isolation systems has led to the development of foundation structures capable of attenuating ground vibrations caused by transportation systems or industrial activities. Such systems are particularly relevant for buildings located near railways, highways, or industrial zones where vibration control is essential for occupant comfort and structural durability.

These case studies highlight the multifunctional capabilities of mechanical metamaterials in architectural engineering. From adaptive façades and acoustic panels to seismic protection and lightweight structural frameworks, metamaterials offer innovative solutions that address multiple challenges simultaneously.

As fabrication technologies such as additive manufacturing, digital design tools, and modular construction methods continue to advance, the practical implementation of metamaterials in architecture is expected to expand further. Future architectural projects may increasingly incorporate metamaterial-based systems to achieve higher performance, sustainability, and adaptability in the built environment.

5.5.1 Case Study: Acoustic Metamaterials in Elbphilharmonie Concert Hall, Germany

One of the most prominent examples of advanced acoustic engineering in architecture is the Elbphilharmonie Hamburg. The concert hall incorporates approximately 10,000 uniquely shaped acoustic panels known as the *White Skin*, designed using parametric algorithms to control sound reflections throughout the performance space.

These panels contain microscopic cavities and irregular surface geometries, functioning similarly to acoustic metamaterials by scattering and diffusing sound waves. The engineered surfaces help distribute sound evenly across the auditorium, ensuring optimal acoustic performance for both performers and audiences.

- Key metamaterial-inspired features include:
- Diffusive microstructured acoustic panels
- Controlled sound reflection and scattering
- Uniform sound distribution in large interior spaces.

This project demonstrates how architected surfaces with micro-scale geometries can significantly improve acoustic performance in architectural environments.

5.5.2 Case Study: Auxetic Metamaterial Façade Systems—ICD/ITKE Research Pavilion, Germany

The ICD/ITKE Research Pavilion developed by the University of Stuttgart explores the application of auxetic metamaterial structures in architectural design.

Auxetic structures exhibit a negative Poisson's ratio, meaning they expand laterally when stretched. In this pavilion, researchers implemented lattice-based architectural elements inspired by metamaterial mechanics, allowing the structure to respond dynamically to mechanical forces.

The pavilion uses lightweight composite materials combined with computationally designed cellular geometries, enabling:

- Adaptive structural deformation
- High strength-to-weight ratio
- Efficient load distribution.

This project illustrates how metamaterial-inspired lattice frameworks can lead to innovative architectural structures with improved structural performance and design flexibility.

5.5.3 Case Study: Metamaterial-Based Thermal Insulation – ETH Zurich Experimental Building Systems

Researchers at ETH Zurich have developed experimental metamaterial-based building insulation panels designed to improve thermal performance in building envelopes.

These panels use periodic cellular structures and porous lattice architectures that significantly reduce heat transfer by creating tortuous pathways for thermal conduction. The air trapped within the cellular structures acts as a natural insulator, enhancing the overall thermal resistance of the wall systems.

Applications demonstrated in experimental building prototypes include:

- High-performance façade insulation panels
- Passive temperature regulation systems
- Lightweight building envelope components.

These studies show how metamaterial-inspired architectures can improve energy efficiency and sustainable building design.

5.5.4 Case Study: Seismic Metamaterial Barriers – Grenoble Seismic Shield Experiment, France

A ground-breaking field experiment conducted near Grenoble demonstrated the concept of seismic metamaterials for earthquake protection. The project involved installing periodic arrays of boreholes in the ground around a test site to manipulate the propagation of seismic waves.

The structured ground array created seismic wave bandgaps, preventing certain frequencies of seismic vibrations from propagating through the protected region. As a result, the energy of the incoming seismic waves was significantly reduced within the protected area.

Key features of this experiment include:

- Periodic underground structures acting as seismic metamaterials
- Attenuation and redirection of seismic waves
- Potential protection for building foundations.

This study demonstrates the feasibility of using metamaterial-inspired ground structures for earthquake-resilient architectural engineering.

5.5.5 Case Study: Metamaterial-Inspired Structural Lattices—MX3D Steel Bridge, Netherlands

The MX3D Bridge in Amsterdam represents a pioneering example of architected lattice structures enabled by advanced manufacturing. Developed by MX3D, the bridge was fabricated using robotic metal 3D printing.

Although not strictly a metamaterial structure, the bridge employs optimized lattice-like internal geometries and topology optimization techniques, which are closely related to metamaterial design principles.

Important features include:

- Lightweight optimized structural lattice
- Reduced material usage with high strength
- Advanced computational structural design.

This project highlights how metamaterial-inspired structural frameworks can enable efficient and sustainable construction through digital fabrication technologies.

5.6 CHALLENGES AND FUTURE OPPORTUNITIES IN ARCHITECTURAL APPLICATIONS

The integration of mechanical metamaterials into architectural engineering presents significant potential for improving structural efficiency, energy performance, acoustic control, and sustainability. However, despite their promising capabilities, the practical implementation of metamaterials in building systems still faces several technical, economic, and design challenges. At the same time, ongoing research and advancements in computational design and manufacturing technologies are creating new opportunities for their broader adoption in architecture.

One of the major challenges in architectural applications of metamaterials is manufacturing complexity. Many metamaterial structures rely on intricate lattice geometries and periodic microstructures that are difficult to fabricate using traditional construction methods. Producing these complex structures at large scales suitable for buildings requires advanced manufacturing techniques such as additive manufacturing, robotic fabrication, and digital construction technologies, which are still evolving and can be costly.

Another challenge is scalability and structural reliability. While metamaterials have demonstrated exceptional mechanical and functional properties in laboratory-scale experiments, translating these properties to full-scale architectural structures requires extensive testing and validation. Engineers must ensure that metamaterial-based components can withstand long-term environmental exposure, mechanical loads, and dynamic stresses associated with real-world building conditions.

Material durability and environmental resistance also present important considerations. Architectural structures must endure varying weather conditions, including temperature fluctuations, moisture, ultraviolet radiation, and mechanical wear. Designing metamaterials that maintain their functional properties over long periods without degradation remains an active area of research.

Cost is another factor that influences the adoption of metamaterials in construction. Advanced fabrication techniques and specialized materials can increase the initial cost of metamaterial-based building components. However, the long-term benefits associated with energy efficiency, reduced material consumption, and multi-functional performance may offset these costs over the lifecycle of a building.

Despite these challenges, several emerging technologies are creating new opportunities for the implementation of metamaterials in architecture. Advances in computational design tools, parametric modeling, and topology optimization allow architects and engineers to design complex metamaterial structures with precise control over mechanical and functional properties. These digital design methods enable the optimization of structural performance while minimizing material usage.

Another promising development is the increasing use of additive manufacturing and robotic construction techniques, which make it possible to fabricate complex lattice structures at larger scales. These technologies allow for the precise manufacturing of architected materials that were previously difficult or impossible to produce using conventional methods.

Future research is also focusing on smart and adaptive metamaterials that can respond dynamically to environmental stimuli such as temperature, stress, or electromagnetic fields. Such materials could enable buildings that automatically adjust their structural behavior, thermal performance, or acoustic properties in response to changing conditions.

In addition, the integration of metamaterials with smart building technologies and sensor networks opens possibilities for intelligent architectural systems capable of monitoring structural performance and optimizing building operation in real time. These systems could improve energy efficiency, safety, and occupant comfort while extending the lifespan of building structures.

Overall, while the widespread application of mechanical metamaterials in architecture is still in its early stages, ongoing research and technological advancements are rapidly expanding their potential. As fabrication technologies mature and design methodologies evolve, metamaterials are expected to play an increasingly important role in the development of innovative, sustainable, and high-performance architectural systems in the future.

5.7 SUMMARY

This chapter explored the growing role of mechanical metamaterials in architectural engineering, highlighting how engineered material architectures can significantly enhance building performance. Unlike conventional materials whose properties depend primarily on composition, mechanical metamaterials derive their unique capabilities from carefully designed internal geometries and periodic structural arrangements. These architected structures allow buildings to achieve improved performance in areas such as thermal insulation, acoustic control, structural efficiency, and sustainability.

The chapter first discussed the thermal and acoustic applications of metamaterials in architectural systems. Metamaterial-based insulation structures with cellular and lattice architectures can effectively reduce heat transfer by creating tortuous thermal conduction paths. These materials contribute to passive temperature control and improved energy efficiency in building envelopes. Similarly, acoustic metamaterials utilize resonant cavities, periodic structures, and engineered micro-geometries to manipulate sound waves, enabling effective soundproofing and noise reduction in interior spaces and urban buildings.

The integration of metamaterials into architectural elements such as walls, roofs, ceilings, and building envelopes was also examined. These systems allow for multifunctional building components capable of providing thermal regulation, acoustic attenuation, and lightweight structural support simultaneously. Such multifunctionality reduces the need for multiple material layers and contributes to more efficient and sustainable building designs.

The chapter further explored structural applications of mechanical metamaterials, including lightweight domes, lattice-based architectural frameworks, and earthquake-resilient structural systems. Architected lattice structures offer high strength-to-weight ratios and efficient load distribution, making them suitable for large-span structures and advanced roofing systems. Additionally, seismic metamaterials and vibration isolation systems can attenuate ground vibrations and seismic waves, improving the resilience and durability of buildings in earthquake-prone areas.

Another important focus of the chapter was the integration of metamaterials with smart and sustainable building systems. Metamaterial-based building envelopes and smart façades can regulate heat transfer, airflow, and solar radiation while improving energy efficiency. These adaptive systems, when combined with sensors and building automation technologies, enable the development of intelligent architectural systems capable of responding to changing environmental conditions.

The chapter also presented several case studies demonstrating the practical application of metamaterials in architectural engineering, including acoustic systems in performance halls, auxetic façade structures, seismic wave mitigation experiments, and lightweight lattice-based structural designs. These examples illustrate the growing potential of metamaterials to transform modern architecture.

Finally, the chapter discussed current challenges and future opportunities associated with the adoption of metamaterials in building systems. Challenges such as manufacturing complexity, scalability, cost, and long-term durability remain areas of ongoing research. However, advances in computational design, digital fabrication, additive manufacturing, and smart material technologies are expected to accelerate the practical implementation of metamaterials in architecture.

REVIEW QUESTIONS

1. What are mechanical metamaterials, and how do they differ from conventional construction materials?
2. Explain the significance of engineered geometry in mechanical metamaterials.
3. Why are metamaterials considered promising for architectural applications?
4. Describe the role of mechanical metamaterials in modern architectural engineering.
5. How do metamaterials improve building performance compared to traditional materials?
6. What is thermal insulation and how can metamaterials enhance it?
7. Explain how cellular metamaterial structures reduce heat transfer.
8. What are the advantages of metamaterial-based thermal insulation systems?
9. Describe the concept of passive temperature control in buildings.
10. How can metamaterial-based building envelopes improve energy efficiency?
11. What are acoustic metamaterials, and how do they control sound propagation?
12. Explain the concept of acoustic bandgaps in metamaterials.
13. How do resonant cavities in metamaterials contribute to sound absorption?
14. Describe the application of acoustic metamaterials in building interiors.
15. How can metamaterials be used to reduce noise in urban buildings?
16. How can metamaterials be integrated into building walls and façades?

17 Discuss the role of metamaterials in roof insulation systems.
18 Explain how metamaterials improve the performance of interior architectural elements.
19 What advantages do metamaterial-based architectural panels offer?
20 How can metamaterials contribute to multifunctional building components?
21 What are the benefits of metamaterial-based lattice structures in architecture?
22 Explain the concept of lightweight architectural frames using metamaterials.
23 How do metamaterials improve load distribution in large-span structures?
24 Describe the application of metamaterials in dome and roof structures.
25 What is the significance of auxetic structures in architectural design?
26 What are seismic metamaterials and how do they protect buildings from earthquakes?
27 Explain how metamaterials create seismic wave bandgaps.
28 Describe the role of metamaterials in vibration isolation systems.
29 How can metamaterial-based foundations reduce ground vibration effects?
30 Why are vibration isolation systems important in building design?
31 What are smart façades, and how do metamaterials contribute to their development?
32 Explain how metamaterials support adaptive architectural systems.
33 How can metamaterials improve the sustainability of buildings?
34 Describe the integration of metamaterials with smart building technologies.
35 What role do metamaterials play in energy-efficient building envelopes?
36 How do metamaterials contribute to lightweight construction?
37 Explain how metamaterials reduce material consumption in buildings.
38 What are the environmental benefits of metamaterial-based construction?
39 How can metamaterials enhance the durability and efficiency of building structures?
40 Discuss the importance of multifunctionality in metamaterial-based building systems.
41 Describe an example of acoustic metamaterials used in architectural design.
42 Explain the use of auxetic structures in architectural façades.
43 What is the significance of seismic metamaterial experiments for earthquake protection?
44 Discuss the role of metamaterial-inspired lattice structures in modern architecture.
45 How have experimental projects demonstrated the potential of metamaterials in building engineering?
46 What are the main challenges in implementing metamaterials in architectural construction?
47 Why is manufacturing complexity a challenge for large-scale metamaterial applications?
48 How can additive manufacturing help overcome metamaterial fabrication challenges?
49 What future opportunities exist for metamaterials in smart and sustainable architecture?
50 How might metamaterials influence the future design of high-performance buildings?

6 Future Trends and Challenges in Mechanical Metamaterials

6.1 INTRODUCTION

Mechanical metamaterials represent a rapidly evolving field of materials science and engineering, characterized by engineered architectures that produce unconventional mechanical properties not typically found in natural materials. Unlike traditional materials whose properties are primarily determined by chemical composition, metamaterials derive their behavior from carefully designed geometric structures at micro- or macro-scales. Over the past decade, significant advances in computational design, fabrication technologies, and interdisciplinary research have accelerated the development of innovative metamaterial systems, leading to numerous emerging trends in this field.

One of the most important emerging trends is the development of programmable and tunable metamaterials. These materials are designed to alter their mechanical properties, such as stiffness, shape, and deformation behavior, in response to external stimuli, including temperature, magnetic fields, mechanical stress, or electrical signals. Programmable metamaterials allow engineers to design structures with adaptive responses, making them suitable for applications in robotics, aerospace systems, and smart infrastructure.

Another significant trend is the advancement of multifunctional metamaterials. Modern metamaterial designs increasingly combine multiple physical properties within a single material architecture. For example, a metamaterial structure may simultaneously provide mechanical strength, thermal insulation, acoustic damping, and vibration control. This multifunctionality reduces the need for multiple separate material systems, improving efficiency in engineering and architectural applications (Gariya et al., 2026).

The rise of additive manufacturing and digital fabrication technologies has also played a crucial role in expanding metamaterial research. Techniques such as 3D printing, laser-based fabrication, and robotic manufacturing enable the production of complex lattice structures and hierarchical architectures that were previously impossible to produce with conventional methods. These fabrication methods enable precise control over internal geometries, which is essential for achieving the desired mechanical properties of metamaterials.

Another emerging direction involves topology-optimized and bio-inspired metamaterials. Researchers are increasingly drawing inspiration from natural structures such as bone, honeycombs, and plant tissues to design efficient cellular architectures.

DOI: 10.1201/9781003783671-6

These designs often exhibit high strength-to-weight ratios, enhanced energy absorption capabilities, and improved structural resilience.

Recent developments have also focused on mechanical metamaterials with extreme properties, including negative Poisson's ratio (auxetic materials), negative stiffness, and programmable deformation patterns. Such materials exhibit unique mechanical behaviors that enable applications in impact protection, deployable structures, and flexible mechanical systems.

Furthermore, the integration of metamaterials with smart materials and sensor technologies is opening new opportunities for intelligent engineering systems. Smart metamaterials that can sense environmental changes and respond accordingly are being explored for use in adaptive structures, soft robotics, aerospace systems, and next-generation architectural components.

As research continues to advance, mechanical metamaterials are expected to play an increasingly important role in addressing modern engineering challenges related to lightweight design, energy efficiency, structural resilience, and multifunctional performance. The exploration of emerging trends in this field provides valuable insights into how metamaterials may shape the future of materials engineering and technological innovation.

6.2 ADVANCED COMPUTATIONAL DESIGN AND OPTIMIZATION

The development of mechanical metamaterials relies heavily on advanced computational design and optimization techniques to create complex architectures with tailored mechanical properties. Because the performance of metamaterials is primarily governed by their internal geometry and structural arrangement, computational tools play a critical role in designing, analyzing, and optimizing these architected materials. Advances in computational modeling, simulation, and algorithm-based optimization have enabled researchers and engineers to design metamaterials with precisely controlled mechanical behavior and multifunctional capabilities.

One of the key approaches used in computational design is topology optimization. This method determines the optimal distribution of material within a given design space while satisfying specific performance requirements such as stiffness, strength, or weight reduction. In metamaterial design, topology optimization allows engineers to generate complex lattice structures and cellular architectures that maximize structural efficiency. By iteratively refining the material layout through numerical algorithms, designers can create lightweight structures with high mechanical performance.

Another important computational tool is parametric modeling, which enables the systematic variation of geometric parameters within a metamaterial structure. Parameters such as cell size, strut thickness, lattice orientation, and periodicity can be adjusted to study their effects on mechanical properties. Parametric design tools, often integrated with computer-aided design (CAD) software, allow researchers to quickly generate and evaluate multiple design configurations, facilitating rapid exploration of design alternatives.

Finite element analysis (FEA) is widely used to simulate the mechanical behavior of metamaterial structures under various loading conditions. FEA models can predict how complex lattice structures respond to forces, vibrations, thermal loads, and dynamic impacts. Through these simulations, engineers can evaluate stress distribution, deformation patterns, and failure mechanisms before physical prototypes are fabricated. This approach significantly reduces development time and improves design reliability.

Another emerging computational approach is machine learning and artificial intelligence (AI)-based design optimization. Machine learning algorithms can analyze large datasets generated from simulations or experiments to identify patterns and relationships between structural geometry and material performance. These algorithms can then guide the design of new metamaterial architectures with optimized properties. AI-driven design methods enable faster discovery of innovative structures that may be difficult to identify using traditional design approaches.

Multiscale modeling techniques are also becoming increasingly important in metamaterial research. Since metamaterials often exhibit behavior influenced by features at different length scales from microscopic structural elements to macroscopic structural performance, multiscale computational models are used to link these scales. Such models allow engineers to predict how microstructural changes affect the overall mechanical behavior of a structure.

Another important aspect of computational design is the use of digital fabrication simulations. These simulations help ensure that the designed metamaterial structures can be manufactured using techniques such as additive manufacturing, laser cutting, or robotic fabrication. By incorporating manufacturing constraints into the computational design process, engineers can develop fabrication-ready metamaterial structures that maintain their intended performance.

6.2.1 Topology Optimization Techniques

Topology optimization is a powerful computational design approach used to determine the optimal material distribution within a given design domain while satisfying specified performance objectives and constraints. In the context of mechanical metamaterials, topology optimization plays a critical role in designing architected lattice structures and cellular geometries that exhibit enhanced mechanical properties such as high stiffness-to-weight ratios, improved energy absorption, and tailored deformation behavior.

Unlike traditional structural optimization methods that only adjust geometric parameters or material dimensions, topology optimization allows the entire structural layout to evolve during the design process. This means that material can be strategically added or removed within the design space to achieve the best possible performance according to predefined objectives such as minimizing weight, maximizing stiffness, or improving thermal and acoustic performance.

One widely used method in topology optimization is the solid isotropic material with penalization (SIMP) method. In this approach, the design domain is divided into numerous finite elements, each assigned a material density variable ranging from zero (void) to one (solid material). Through iterative numerical calculations, the algorithm adjusts the density distribution across the domain to identify the optimal structural configuration. The SIMP method is widely used because it provides efficient solutions for complex structural optimization problems (Zheng et al., 2020).

Another important technique is the level-set method, which represents structural boundaries using mathematical functions. Instead of assigning density values to elements, this method describes the geometry of the structure through evolving contours. The level-set method allows for smooth boundary representation and is particularly useful for designing metamaterial architectures where precise control of structural geometry is required.

The evolutionary structural optimization (ESO) method is another approach used in metamaterial design. In this technique, inefficient material regions that contribute little to structural performance are gradually removed from the design domain. The process continues iteratively until the optimized structure is achieved. A related method, bi-directional evolutionary structural optimization (BESO), allows both material removal and addition during optimization, leading to more refined structural solutions.

Topology optimization is particularly valuable in the design of periodic metamaterial unit cells, which form the building blocks of larger metamaterial structures. By optimizing the geometry of these unit cells, engineers can achieve targeted mechanical properties such as negative Poisson's ratio, tunable stiffness, or enhanced energy absorption. These optimized unit cells can then be repeated to create large-scale metamaterial systems used in engineering and architectural applications.

Modern topology optimization techniques often integrate with FEA to evaluate structural performance under various loading conditions. The optimization algorithm uses simulation results to iteratively modify the material distribution until the design objectives are satisfied.

Advancements in computational power, parametric design tools, and additive manufacturing technologies have significantly enhanced the applicability of topology optimization in metamaterial design. Complex optimized structures that were previously difficult to fabricate can now be manufactured using 3D printing and digital fabrication techniques.

6.2.2 Generative and Algorithmic Design Methods

Generative and algorithmic design methods have become essential tools in the development of advanced mechanical metamaterials, enabling the creation of highly complex architectures that would be difficult to design using traditional approaches. These computational design techniques use mathematical algorithms, parametric modeling, and automated design processes to generate optimized structural configurations based on predefined performance criteria. By integrating these methods with simulation and optimization tools, researchers can efficiently explore a large range of metamaterial geometries and identify structures with desirable mechanical properties.

Generative design refers to a computational process in which a design algorithm automatically produces multiple structural solutions based on specified constraints and objectives. Designers provide input parameters such as material type, load conditions, geometric boundaries, and performance requirements, and the algorithm generates numerous design alternatives. These solutions are then evaluated using computational simulations, allowing engineers to select the most efficient configuration for the intended application (Wang et al., 2020).

Algorithmic design, on the other hand, focuses on the use of rule-based computational procedures to define and control the geometry of metamaterial structures. In this approach, mathematical relationships and logical operations are used to generate patterns, lattice frameworks, and cellular architectures. Algorithmic methods enable precise control over parameters such as unit-cell geometry, periodicity, orientation, and hierarchical structure, which are critical factors in determining the mechanical behavior of metamaterials.

One important advantage of generative and algorithmic design methods is their ability to create highly complex lattice structures and hierarchical metamaterial architectures. These structures often exhibit superior performance characteristics, including enhanced stiffness-to-weight ratios, improved energy absorption capabilities, and tunable deformation responses. Because the design process is automated, it allows researchers to explore design spaces that would be difficult or time-consuming to investigate manually.

Another key feature of these design approaches is their integration with parametric modeling platforms. Parametric design tools allow designers to adjust geometric parameters dynamically and instantly observe how changes affect structural performance. This capability enables rapid iteration and optimization of metamaterial structures.

Generative design methods are often combined with simulation-driven optimization techniques such as FEA. Each generated design can be analyzed under various loading conditions to evaluate its structural performance. The algorithm can then refine the design iteratively until the optimal configuration is achieved.

In recent years, the integration of AI and machine learning algorithms with generative design has further enhanced the development of metamaterials. Machine learning models can analyze large datasets of metamaterial designs and predict optimal structural configurations based on performance objectives. This approach accelerates the discovery of innovative metamaterial architectures.

Generative and algorithmic design methods are also closely linked with digital fabrication technologies, including additive manufacturing and robotic construction. These fabrication techniques enable the realization of complex geometries generated through computational design processes, allowing designers to translate digital metamaterial architectures into physical structures.

6.2.3 Finite Element Modeling of Metamaterial Architectures

Finite element modeling (FEM) is one of the most important computational tools used in the design and analysis of mechanical metamaterial architectures. Since the performance of metamaterials is largely determined by their complex internal geometries and lattice structures, analytical solutions are often difficult to obtain. FEM provides a numerical approach that allows engineers to simulate and predict the mechanical behavior of metamaterials under different loading and environmental conditions (Zhang et al., 2024).

The finite element method works by dividing a complex structure into a large number of small interconnected elements, collectively known as a mesh. Each element is governed by mathematical equations that describe its mechanical behavior. By assembling these elements and solving the system of equations, FEM can predict important structural responses such as stress distribution, deformation patterns, strain localization, and failure mechanisms within metamaterial structures.

In metamaterial research, FEM is commonly used to study the behavior of periodic unit cells, which serve as the fundamental building blocks of larger metamaterial structures. Engineers often analyze a single representative unit cell with appropriate boundary conditions to determine the effective mechanical properties of the overall material. These properties may include effective stiffness, Poisson's ratio, energy absorption capacity, and elastic modulus.

Another important application of FEM in metamaterial design is the simulation of lattice structures and cellular architectures under different loading conditions such as compression, tension, bending, and shear. By performing these simulations, researchers can understand how structural geometry influences mechanical performance and identify optimal configurations for specific applications.

FEM is also widely used for analyzing dynamic behavior and vibration characteristics of metamaterials. Metamaterial structures often exhibit unique wave propagation properties such as bandgaps and vibration attenuation, which are important for applications in acoustic control, vibration isolation, and seismic protection. FEM simulations allow engineers to investigate how elastic waves interact with periodic structures and how these interactions affect the overall mechanical response.

Another significant advantage of FEM is its ability to incorporate material nonlinearities and complex boundary conditions. Metamaterials may undergo large deformations or exhibit nonlinear mechanical behavior due to their geometric design. Finite element models can account for these nonlinear effects, enabling more accurate prediction of real-world performance.

In addition to structural analysis, FEM is often integrated with optimization and parametric design techniques. Engineers can modify geometric parameters, such as cell size, strut thickness, and lattice orientation, and evaluate their effects on mechanical performance through repeated simulations. This process helps identify optimized metamaterial architectures with improved functionality.

Modern FEM is supported by advanced computational software packages such as ANSYS, ABAQUS, COMSOL Multiphysics, and LS-DYNA, which provide specialized tools for analyzing complex structures. These platforms allow researchers to simulate both static and dynamic behavior of metamaterials with high accuracy.

6.2.4 Machine Learning and Artificial Intelligence in Metamaterial Design

Machine learning (ML) and AI are increasingly being integrated into the design and development of mechanical metamaterials, providing powerful tools for discovering new architectures and optimizing material performance. Traditional metamaterial design methods often rely on trial-and-error simulations and computational optimization, which can be time-consuming and computationally expensive. ML and AI techniques offer a more efficient approach by enabling data-driven design and predictive modeling (Cerniauskas et al., 2024).

ML algorithms can analyze large datasets generated from numerical simulations, experimental measurements, and computational design studies to identify relationships between metamaterial geometry and mechanical properties. By learning from these datasets, ML models can predict how changes in structural parameters—such as lattice geometry, unit-cell configuration, and material distribution—affect the overall mechanical behavior of the metamaterial.

One common application of ML in metamaterial design is property prediction. Supervised learning algorithms can be trained on simulation data to predict mechanical properties such as elastic modulus, Poisson's ratio, strength, and energy absorption capacity. Once trained, these models can rapidly evaluate new metamaterial designs without the need for extensive numerical simulations, significantly accelerating the design process.

Another important application is inverse design, where AI algorithms determine the optimal structural configuration required to achieve specific mechanical properties. In this approach, designers specify target properties such as desired stiffness, negative Poisson's ratio, or vibration attenuation characteristics, and the algorithm generates potential metamaterial architectures that meet these requirements. Inverse design methods enable the discovery of innovative structures that may not be intuitive through traditional design approaches.

Generative ML models, including deep learning networks and generative adversarial networks (GANs), are also being used to create new metamaterial geometries automatically. These models can explore large design spaces and generate complex lattice structures that satisfy performance constraints. By combining generative models with optimization techniques, researchers can efficiently identify high-performance metamaterial architectures.

Reinforcement learning is another AI-based approach that has shown promise in metamaterial design. In reinforcement learning frameworks, an algorithm iteratively modifies the geometry of a metamaterial structure and evaluates its performance using simulation feedback. Through repeated iterations, the system learns which design strategies lead to improved material properties, gradually converging toward optimized solutions.

ML techniques are also being integrated with FEA and topology optimization. In such hybrid approaches, AI models assist in guiding the optimization process by predicting promising design regions within the search space. This significantly reduces computational time while maintaining high design accuracy.

Another advantage of AI-based design is its ability to handle multifunctional metamaterial optimization. Mechanical metamaterials often need to satisfy multiple performance criteria simultaneously, such as structural strength, thermal insulation, and acoustic attenuation. ML algorithms can efficiently analyze complex relationships among these parameters and identify optimal trade-offs between competing design objectives.

Despite these advantages, challenges remain in applying ML to metamaterial design. High-quality datasets are required for training accurate models, and the interpretability of AI-generated designs can sometimes be limited. Nevertheless, ongoing research in explainable AI, hybrid modeling, and physics-informed ML is addressing these challenges.

Table 6.1 presents a comparative overview of major computational design and optimization techniques used in the development of mechanical metamaterials.

TABLE 6.1
Comparative Study of Advanced Computational Design and Optimization Methods for Mechanical Metamaterials

Design Method	Principle/Approach	Key Features	Advantages	Limitations	Example Application in Metamaterials
Topology optimization	Determines the optimal distribution of material within a design domain to achieve specified performance objectives.	Uses algorithms such as SIMP, BESO, and level-set methods to iteratively refine structural layouts.	Produces lightweight structures with high stiffness-to-weight ratios and efficient load distribution.	Computationally intensive for large or highly detailed structures.	Optimization of lattice unit cells for lightweight aerospace and structural components.
Parametric modeling	Controls metamaterial geometry using adjustable parameters such as cell size, strut thickness, and orientation.	Enables systematic variation of structural parameters within CAD-based models.	Allows rapid design modification and efficient exploration of design alternatives.	May not always provide globally optimal designs without additional optimization methods.	Design of auxetic metamaterial unit cells with tunable mechanical properties.
Finite element analysis (FEA)	Uses numerical simulation to analyze stress, deformation, vibration, and failure behavior of metamaterial structures.	Divides complex geometries into small finite elements for detailed mechanical analysis.	Provides accurate prediction of mechanical response under various loading conditions.	Requires significant computational resources for complex metamaterial geometries.	Simulation of mechanical performance of cellular and lattice metamaterial structures.
Generative design algorithms	Automatically generates multiple structural designs based on predefined constraints and objectives.	Uses rule-based algorithms and optimization processes to explore large design spaces.	Enables discovery of innovative metamaterial architectures and complex geometries.	Generated designs often require engineering validation and refinement.	Development of complex cellular metamaterial frameworks for structural applications.
Machine learning-based optimization	Utilizes data-driven models to predict and optimize metamaterial properties and geometries.	Employs neural networks, deep learning, and AI algorithms for design prediction.	Accelerates design discovery and reduces computational time for property evaluation.	Requires large datasets for accurate training and validation.	AI-assisted design of metamaterials with negative Poisson's ratio and tailored stiffness.
Multiscale modeling	Connects microstructural geometry with macroscopic material behavior across multiple length scales.	Combines microstructure simulation with structural-level analysis.	Captures realistic mechanical performance of metamaterials across different scales.	Modeling complexity increases due to scale coupling and computational requirements.	Prediction of effective mechanical properties of hierarchical lattice metamaterials.

Each method offers unique capabilities for designing and analyzing complex architected structures. The integration of these approaches enables the creation of high-performance metamaterials with optimized mechanical, thermal, and multifunctional properties suitable for advanced engineering and architectural applications.

6.3 PROGRAMMABLE AND ADAPTIVE METAMATERIALS

Programmable and adaptive metamaterials represent a significant advancement in the field of mechanical metamaterials, enabling structures to change their mechanical behavior or physical properties in response to external stimuli. Unlike traditional materials with fixed mechanical characteristics, these metamaterials are designed with reconfigurable architectures and smart material components that allow them to adapt to varying environmental conditions or operational requirements.

Programmable metamaterials are engineered so that their mechanical properties, deformation patterns, or structural configurations can be controlled through external inputs. These inputs may include mechanical forces, temperature changes, magnetic or electric fields, and pneumatic or hydraulic actuation. By carefully designing the internal geometry and actuation mechanisms, engineers can create metamaterials whose behavior can be programmed to perform specific functions (Wang et al., 2025).

One important characteristic of programmable metamaterials is their ability to exhibit multiple stable configurations, often referred to as multistability. Structures with multistable behavior can switch between different shapes or mechanical states without continuous energy input. This property is particularly useful for applications such as deployable structures, adaptive architectural components, and morphing aerospace structures.

Adaptive metamaterials, on the other hand, are capable of automatically responding to environmental changes without direct external control. These materials often incorporate smart materials such as shape memory alloys, piezoelectric materials, magnetorheological materials, or responsive polymers. When exposed to specific environmental stimuli, these materials trigger structural changes within the metamaterial architecture, allowing the system to adjust its mechanical properties dynamically.

One of the key advantages of programmable and adaptive metamaterials is their ability to enable tunable mechanical properties. For example, a metamaterial structure may be designed to change its stiffness, damping characteristics, or vibration response depending on external conditions. This capability is particularly valuable in applications involving vibration control, impact protection, and structural adaptability.

Another emerging concept is the development of reconfigurable metamaterial lattices, where the arrangement of unit cells can be altered to modify the overall mechanical behavior of the structure. Such systems may use mechanical linkages, modular components, or embedded actuators to rearrange internal structural patterns. This allows the material to adapt its structural performance according to functional requirements.

Programmable metamaterials also play a crucial role in the development of soft robotics and deployable engineering systems. In these applications, metamaterial structures can undergo controlled shape transformations, enabling flexible movement and adaptive mechanical responses. Similar concepts are being explored in architecture, where adaptive metamaterial panels could regulate airflow, shading, or structural stiffness in response to environmental conditions.

Advances in additive manufacturing and digital fabrication technologies have significantly accelerated the development of programmable metamaterials. These fabrication techniques allow the creation of intricate geometries and the integration of smart materials directly within metamaterial architectures. As a result, researchers can design complex structures capable of controlled deformation and dynamic functionality.

Despite these advancements, challenges remain in the practical implementation of programmable and adaptive metamaterials. Issues such as fabrication complexity, material durability, and system integration must be addressed before widespread adoption becomes feasible. Nevertheless, ongoing research in smart materials, computational design, and advanced manufacturing continues to expand the possibilities of these systems.

6.3.1 Smart Materials Integration

The integration of smart materials into mechanical metamaterial architectures has opened new possibilities for developing adaptive, responsive, and multifunctional material systems. Smart materials possess the ability to sense and respond to external stimuli such as temperature, mechanical stress, electric or magnetic fields, light, and chemical changes. When combined with the engineered geometries of metamaterials, these materials enable the creation of structures capable of dynamic property modulation and real-time adaptability.

Mechanical metamaterials typically derive their unique properties from their geometric configuration and periodic structural arrangement. However, by embedding smart materials within these architectures, it becomes possible to actively control or tune the metamaterial's behavior. This integration allows the structure to modify properties such as stiffness, damping capacity, shape, and vibration response depending on environmental conditions or external commands (Zhang et al., 2025).

One widely studied smart material used in metamaterial systems is the shape memory alloy (SMA). Shape memory alloys can recover their original shape after deformation when exposed to a specific temperature change. When integrated into metamaterial frameworks, SMAs enable structures that can change configuration or stiffness in response to thermal stimuli, making them useful in deployable structures, aerospace components, and adaptive architectural elements.

Another important category is piezoelectric materials, which generate an electric charge when subjected to mechanical stress and conversely deform when exposed to an electric field. Piezoelectric components embedded in metamaterial structures allow for active vibration control and sensing capabilities, enabling structures that can monitor mechanical stress and adjust their response accordingly.

Magnetorheological (MR) and electrorheological (ER) materials are also increasingly integrated with metamaterials. These materials can change their mechanical properties such as viscosity or stiffness when exposed to magnetic or electric fields. When used in metamaterial architectures, they provide controllable damping and tunable mechanical responses, which are beneficial for vibration isolation and adaptive structural systems.

Another emerging area involves the use of responsive polymers and hydrogels in metamaterial structures. These materials can undergo significant deformation when exposed to environmental changes such as temperature, humidity, or chemical stimuli. Integrating such materials into metamaterial lattices enables the creation of soft adaptive systems that can change shape or mechanical behavior automatically.

The integration of smart materials with metamaterial architectures also facilitates the development of self-sensing and self-actuating structures. In such systems, embedded sensors detect environmental changes or structural loads, while actuators modify the material configuration or stiffness in response. These capabilities are essential for the development of intelligent structures and smart infrastructure systems.

Advanced fabrication technologies such as multi-material additive manufacturing and digital fabrication play a critical role in enabling smart material integration. These techniques allow different materials with distinct functional properties to be incorporated within a single metamaterial structure, enabling complex multifunctional systems.

Despite the promising potential of smart material integration, several challenges remain. Issues such as material compatibility, durability, power requirements for actuation, and fabrication complexity must be addressed to achieve reliable and scalable implementations.

6.3.2 Shape-Morphing and Reconfigurable Structures

Shape-morphing and reconfigurable structures represent an important class of programmable mechanical metamaterials capable of undergoing controlled geometric transformations in response to external stimuli or internal actuation mechanisms. These structures are designed so that their shape, configuration, or mechanical properties can be dynamically altered, enabling adaptive functionality in various engineering applications. Unlike conventional materials with fixed structural configurations, shape-morphing metamaterials can transform between different structural states while maintaining mechanical integrity.

Shape-morphing behavior in metamaterials is primarily achieved through carefully engineered lattice geometries and responsive material components. The internal architecture of these materials often includes periodic unit cells that allow large deformation without structural failure. These unit cells can be designed to undergo folding, bending, twisting, or expansion, resulting in significant changes in the overall shape of the structure.

One common mechanism used in shape-morphing metamaterials is mechanical instabilities and buckling-based deformation. In such systems, the metamaterial structure is designed so that controlled buckling occurs under specific loading conditions. Instead of leading to structural failure, the buckling behavior produces a predictable change in geometry that allows the structure to transition between different configurations (Dudek et al., 2025).

Another important approach involves the use of multistable structures, which possess multiple stable equilibrium states. Multistable metamaterials can switch between these states when subjected to external forces or environmental stimuli. Once the structure transitions to a new configuration, it can remain stable without continuous energy input. This property is particularly valuable for applications requiring deployable or reconfigurable structures.

Shape morphing can also be achieved by integrating smart materials such as shape memory alloys, piezoelectric materials, or responsive polymers into metamaterial architectures. These materials respond to environmental triggers such as temperature, electric fields, or humidity, enabling the structure to change its geometry autonomously. For example, temperature-induced phase transformations in shape memory alloys can cause metamaterial structures to expand, contract, or bend, resulting in controlled shape transformations.

Reconfigurable metamaterials often incorporate modular lattice architectures or mechanical linkages that allow structural elements to rearrange their positions. In such systems, the geometry of the metamaterial unit cells can be altered to modify properties such as stiffness, load-bearing capacity, or vibration response. This adaptability makes reconfigurable metamaterials particularly useful in applications where structural requirements change over time.

Shape-morphing metamaterials have numerous potential applications across different fields. In aerospace engineering, they are used for deployable structures such as morphing wings and satellite components. In architecture, adaptive metamaterial panels can regulate airflow, shading, or structural stiffness in response to environmental conditions. In robotics, flexible metamaterial structures enable the development of soft robots capable of complex movements and adaptive interactions with their surroundings.

Advances in additive manufacturing and digital fabrication technologies have greatly facilitated the development of shape-morphing metamaterials. These technologies allow precise fabrication of intricate lattice geometries and enable the integration of multiple materials with different functional properties.

Despite significant progress, challenges remain in the design and implementation of shape-morphing metamaterials. Issues related to structural stability, durability under repeated transformations, and reliable actuation mechanisms must be addressed for large-scale practical applications.

6.3.3 Stimuli-Responsive Metamaterials

Stimuli-responsive metamaterials are an advanced class of adaptive mechanical metamaterials that can alter their structural configuration, mechanical properties, or functional behavior in response to external environmental stimuli. These metamaterials combine engineered geometric architectures with responsive materials, enabling dynamic changes in their performance without requiring complex mechanical systems. Such adaptability makes them highly attractive for applications in smart structures, robotics, aerospace systems, and adaptive architectural designs.

Stimuli-responsive metamaterials operate based on the interaction between external triggers and the internal structure of the material. When exposed to specific stimuli such as temperature changes, magnetic fields, electric fields, light, humidity, or chemical signals, the metamaterial undergoes controlled structural modifications. These changes may involve variations in stiffness, shape, damping capacity, or wave propagation characteristics.

One of the most common types of stimuli-responsive metamaterials is thermally responsive metamaterials. These systems often incorporate materials such as shape memory alloys or temperature-sensitive polymers, which change their shape or mechanical properties when heated or cooled. The transformation can cause the metamaterial lattice to expand, contract, or reconfigure, allowing the structure to adapt to changing thermal conditions (Li et al., 2022).

Another important category includes electrically responsive metamaterials, which use materials such as piezoelectric or electroactive polymers. These materials deform when subjected to an electric field, enabling the metamaterial to adjust its mechanical response or vibration characteristics. Electrically responsive systems are particularly useful in applications requiring active control of structural vibrations or dynamic mechanical tuning.

Magnetically responsive metamaterials represent another emerging area of research. These systems integrate materials such as magnetorheological fluids or magnetic particles within metamaterial structures. When exposed to a magnetic field, the mechanical properties of these materials change, allowing the metamaterial to exhibit tunable stiffness or controllable damping behavior. Such systems are widely explored for vibration isolation and adaptive mechanical systems.

Humidity-responsive and chemically responsive metamaterials are also gaining attention, particularly in soft material systems. In these materials, responsive polymers or hydrogels absorb moisture or react to chemical signals, leading to changes in volume or mechanical stiffness. This type of behavior is useful in applications involving environmental sensing and soft robotic systems.

One of the most important advantages of stimuli-responsive metamaterials is their ability to enable self-adaptive structures. Instead of relying on external mechanical systems for control, these materials can automatically adjust their properties based on environmental conditions. This capability reduces the complexity of engineering systems and enhances their reliability.

Recent advances in multi-material additive manufacturing and digital fabrication techniques have significantly accelerated the development of stimuli-responsive metamaterials. These technologies allow researchers to integrate responsive materials directly into metamaterial architectures with high precision, enabling the creation of complex structures capable of dynamic behavior.

Despite their promising capabilities, stimuli-responsive metamaterials face several challenges, including material durability, response speed, and long-term stability under repeated environmental cycles. Researchers are actively working to improve these aspects through the development of new responsive materials and optimized structural designs.

6.3.4 Soft Robotic Applications

Soft robotics is an emerging field that focuses on the development of flexible, compliant, and adaptive robotic systems capable of interacting safely with complex and unstructured environments. Mechanical metamaterials have become increasingly important in this field due to their programmable mechanical behavior, tunable stiffness, and ability to undergo large reversible deformations. By integrating metamaterial architectures into soft robotic systems, engineers can design robots that exhibit enhanced flexibility, controllable motion, and multifunctional capabilities.

Unlike traditional rigid robotic systems, soft robots are typically constructed from elastic materials such as polymers, elastomers, hydrogels, and smart materials. These materials allow the robot to deform continuously and adapt its shape during movement. Mechanical metamaterials enhance these properties by providing structured internal geometries that guide deformation patterns and mechanical responses. Through carefully designed lattice architectures and periodic unit cells, metamaterials can control how the robotic structure bends, stretches, twists, or compresses.

One of the key advantages of using metamaterials in soft robotics is the ability to achieve programmable deformation. The internal architecture of metamaterial structures can be designed so that specific regions of the robot deform in predetermined ways when subjected to external forces or internal actuation. This capability enables the creation of robotic systems with controlled motion patterns, such as bending fingers, twisting actuators, or expanding structures.

Another important feature is tunable stiffness. Soft robots often require a balance between flexibility and structural support. Metamaterial-based designs allow engineers to adjust the stiffness of robotic components by modifying the geometry of lattice structures or by incorporating responsive materials. This enables robots to become either soft and flexible for delicate tasks or stiff and stable for load-bearing operations.

Mechanical metamaterials are also widely used in the development of soft robotic actuators. These actuators are responsible for generating movement within the robot. Metamaterial architectures can amplify motion and distribute forces efficiently, allowing actuators to produce complex movements such as gripping, crawling, and shape transformation. For example, metamaterial-based pneumatic actuators can expand or contract when air pressure is applied, enabling smooth and adaptive robotic motion.

Another emerging application involves the development of bio-inspired soft robotic systems. Many biological organisms, such as octopuses, worms, and starfish, exhibit flexible and adaptable movement. Metamaterial architectures can mimic these biological mechanisms by enabling robots to perform continuous deformation and multi-directional movement, allowing them to navigate complex environments.

Soft robots incorporating metamaterials are particularly useful in applications requiring safe human–robot interaction. Their compliant structures reduce the risk of injury during contact with humans, making them suitable for fields such as medical devices, rehabilitation systems, wearable robotics, and assistive technologies.

Advances in additive manufacturing and multi-material fabrication technologies have greatly facilitated the development of metamaterial-based soft robotic systems. These manufacturing techniques enable the precise fabrication of intricate lattice structures and allow the integration of multiple materials with different mechanical properties within a single robotic component.

Despite the promising potential of metamaterials in soft robotics, several challenges remain. Issues related to material durability, actuation efficiency, control complexity, and energy supply must be addressed to achieve reliable and scalable soft robotic systems.

Table 6.2 presents a comparative overview of different programmable and adaptive metamaterial systems used in mechanical metamaterial design. These materials

TABLE 6.2
Comparative Study of Programmable and Adaptive Metamaterials for Mechanical Metamaterials

Metamaterial Type	Working Principle	Key Stimulus/ Control Mechanism	Major Characteristics	Advantages	Example Applications
Shape Memory Metamaterials	Utilize materials that recover their original shape after deformation when exposed to temperature changes.	Thermal stimulus (temperature variation).	Shape recovery, programmable deformation, thermal responsiveness.	Enables deployable and self-reconfiguring structures.	Aerospace deployable structures, adaptive architectural panels.
Piezoelectric Metamaterials	Convert mechanical energy into electrical signals and vice versa to control structural deformation.	Electrical stimulus or mechanical stress.	Active sensing, vibration control, tunable stiffness.	Enables real-time monitoring and active structural control.	Vibration damping systems, smart sensors, adaptive mechanical structures.
Magnetorheological Metamaterials	Contain materials whose mechanical properties change when subjected to magnetic fields.	Magnetic field.	Variable stiffness, controllable damping properties.	Rapid response and adjustable mechanical behavior.	Adaptive vibration isolation systems, automotive suspension components.
Electroactive Polymer Metamaterials	Use electroactive polymers that deform under applied electric fields.	Electrical stimulus.	Large deformation capability, lightweight actuation.	Suitable for flexible and soft robotic structures.	Soft robotic actuators, artificial muscles.
Humidity-Responsive Metamaterials	Incorporate materials that expand or contract based on moisture absorption.	Environmental humidity or chemical stimulus.	Shape morphing, moisture sensitivity, passive adaptability.	Enables self-adaptive environmental response.	Smart ventilation systems, environmental sensors.
Thermoresponsive Polymer Metamaterials	Use polymers that change mechanical behavior with temperature changes.	Temperature stimulus.	Reversible expansion and contraction, tunable stiffness.	Passive adaptation to environmental temperature changes.	Smart building envelopes, adaptive thermal systems.

respond to various external stimuli such as temperature, electric fields, magnetic fields, and humidity, enabling dynamic control of mechanical properties and structural configurations. The integration of such adaptive materials within metamaterial architectures facilitates the development of smart structures, soft robotic systems, adaptive building components, and intelligent engineering systems.

6.4 MULTIFUNCTIONAL METAMATERIAL SYSTEMS

Multifunctional metamaterial systems represent an important advancement in the field of mechanical metamaterials, where a single material architecture is designed to perform multiple functions simultaneously. Traditional engineering materials are typically optimized for a single property, such as strength, thermal insulation, or acoustic control. In contrast, multifunctional metamaterials combine structural, mechanical, thermal, acoustic, and electromagnetic capabilities within a single engineered architecture, enabling more efficient and compact system designs.

The multifunctionality of metamaterials arises primarily from their carefully designed internal geometries and periodic lattice structures. By modifying the geometry, topology, and hierarchical arrangement of unit cells, engineers can tailor different physical properties within the same structure. For example, a metamaterial lattice may provide high structural strength while simultaneously offering vibration damping and thermal insulation.

One key feature of multifunctional metamaterials is their ability to integrate mechanical performance with functional properties. For instance, structural metamaterials used in aerospace or architectural applications can be designed to provide load-bearing support while also controlling heat transfer, acoustic wave propagation, or vibration transmission. This integration reduces the need for multiple separate components, leading to lighter and more efficient systems.

Another important aspect of multifunctional metamaterial systems is the incorporation of smart materials and responsive components. When combined with responsive materials such as piezoelectric elements, shape memory alloys, or electroactive polymers, metamaterials can achieve dynamic multifunctionality. These systems can actively adjust properties such as stiffness, damping, or shape in response to external stimuli, making them suitable for adaptive engineering systems.

Multifunctional metamaterials also enable the development of energy-absorbing and protective structures. Cellular metamaterial architectures can be engineered to absorb impact energy while simultaneously maintaining structural stability and vibration control. Such materials are widely explored for use in automotive safety systems, protective equipment, and aerospace components.

In architectural engineering, multifunctional metamaterials are increasingly used to develop high-performance building systems. For example, metamaterial panels can provide structural support, thermal insulation, acoustic attenuation, and ventilation control within a single architectural component. This capability contributes to the development of energy-efficient and sustainable building designs.

The integration of additive manufacturing technologies has significantly enhanced the development of multifunctional metamaterial systems. Advanced manufacturing methods such as 3D printing and multi-material fabrication allow precise control over internal architectures and enable the combination of materials with different functional properties within a single structure.

Despite their promising capabilities, several challenges remain in the design and implementation of multifunctional metamaterials. Achieving optimal performance for multiple functions simultaneously can be complex, as improving one property may sometimes compromise another. Therefore, advanced computational design, optimization techniques, and experimental validation are necessary to achieve balanced performance.

6.4.1 Coupled Mechanical–Thermal Metamaterials

Coupled mechanical–thermal metamaterials are a class of multifunctional metamaterials designed to simultaneously control mechanical behavior and thermal transport properties through carefully engineered structural architectures. Unlike conventional materials where mechanical and thermal properties are primarily determined by material composition, these metamaterials utilize geometric design and structural topology to achieve coordinated control of both mechanical deformation and heat transfer.

The coupling between mechanical and thermal properties arises from the interaction between structural geometry, material composition, and thermal conduction mechanisms. In many metamaterial architectures, cellular lattices and porous frameworks influence not only the mechanical stiffness and strength but also the effective thermal conductivity of the structure. By adjusting parameters such as unit-cell geometry, strut thickness, and lattice orientation, engineers can tune both the mechanical performance and thermal behavior of the material.

One of the most common design strategies for mechanical–thermal coupling involves the use of lattice-based metamaterial structures. These structures contain interconnected networks of beams or struts that form periodic unit cells. The geometry of these cells determines how forces are distributed through the structure while also influencing the pathways available for heat conduction. As a result, the metamaterial can simultaneously provide mechanical load-bearing capability and thermal insulation or heat management.

Another important approach involves thermally responsive materials integrated within metamaterial architectures. Materials such as shape memory alloys or thermally responsive polymers can change their mechanical stiffness or structural configuration in response to temperature variations. When incorporated into metamaterial frameworks, these materials enable structures that can adapt their mechanical behavior based on thermal conditions.

Coupled mechanical–thermal metamaterials are particularly useful in applications requiring simultaneous structural support and thermal regulation. For example, in aerospace engineering, lightweight metamaterial panels can provide structural strength while also controlling heat transfer from high-temperature environments. Similarly, in architecture, metamaterial-based building components can offer both mechanical stability and thermal insulation, improving overall building energy efficiency.

Another important application is in thermal stress management. Structures exposed to temperature gradients often experience thermal expansion or contraction, which can lead to mechanical stress and potential structural damage. Metamaterial architectures can be designed to accommodate thermal expansion through controlled deformation mechanisms, reducing stress concentrations and improving structural durability.

Recent research has also explored the development of thermal cloaking and directional heat transfer metamaterials. These structures can manipulate the path of heat flow through the material, allowing engineers to guide thermal energy away from sensitive components while maintaining structural integrity.

Advances in additive manufacturing and multi-material fabrication technologies have greatly facilitated the design and production of coupled mechanical–thermal metamaterials. These manufacturing methods allow precise control over internal geometries and enable the integration of materials with different thermal and mechanical properties within a single structure.

Despite these advancements, several challenges remain in the design of coupled mechanical–thermal metamaterials. Achieving optimal performance for both mechanical strength and thermal control requires careful balancing of structural parameters. Additionally, accurately modeling the interaction between mechanical deformation and heat transfer often requires advanced computational simulations and multiscale modeling techniques.

6.4.2 Mechanical–Acoustic Metamaterials

Mechanical–acoustic metamaterials are a class of multifunctional metamaterials designed to control both mechanical structural behavior and acoustic wave propagation through engineered microstructures. These materials combine mechanical strength and stiffness with the ability to manipulate sound waves, enabling advanced solutions for vibration control, sound attenuation, and structural performance in various engineering applications.

Unlike conventional materials where acoustic properties depend mainly on density and elasticity, mechanical–acoustic metamaterials derive their unique behavior from carefully designed periodic structures and resonant elements. The internal geometry of these metamaterials, such as lattice frameworks, resonant cavities, or membrane-based architectures, allows them to interact with acoustic waves in unusual ways, producing effects such as sound absorption, wave redirection, and acoustic bandgaps.

One of the most important characteristics of mechanical–acoustic metamaterials is their ability to create acoustic bandgaps, which are frequency ranges in which sound waves cannot propagate through the material. These bandgaps are produced by the periodic arrangement of structural elements within the metamaterial. When sound waves encounter such periodic structures, destructive interference occurs, preventing wave transmission within certain frequency ranges. This property makes these materials highly effective for noise reduction and vibration isolation applications.

Another key mechanism in mechanical–acoustic metamaterials is local resonance. In this approach, small resonant units embedded within the metamaterial vibrate at specific frequencies when exposed to sound waves. These resonators absorb and dissipate acoustic energy, thereby reducing the amplitude of transmitted sound waves. Local resonance is particularly effective for controlling low-frequency noise, which is often difficult to attenuate using conventional soundproofing materials.

Mechanical–acoustic metamaterials are also designed to maintain structural load-bearing capacity while providing acoustic functionality. Lattice structures and cellular architectures can support mechanical loads while simultaneously acting as sound-damping systems. This multifunctionality is especially valuable in engineering systems where both structural performance and acoustic control are required.

Applications of mechanical–acoustic metamaterials can be found in a wide range of fields. In architectural engineering, these materials are used in building panels, ceilings, and wall systems to improve acoustic comfort and reduce environmental noise. In transportation systems, such as aircraft, automobiles, and trains, mechanical–acoustic metamaterials help reduce engine noise and structural vibrations. In industrial environments, they are used for vibration damping and noise isolation in machinery and equipment.

Recent developments have also explored the integration of mechanical–acoustic metamaterials with adaptive and programmable systems. By incorporating smart materials or adjustable structural elements, these metamaterials can achieve tunable acoustic properties, allowing engineers to control sound transmission dynamically.

Advances in additive manufacturing and digital fabrication techniques have greatly facilitated the development of mechanical–acoustic metamaterials. These technologies enable the fabrication of complex lattice structures and resonant architectures that are difficult to produce using traditional manufacturing methods.

Despite their advantages, challenges remain in the practical implementation of mechanical acoustic metamaterials. Designing structures that simultaneously achieve high mechanical strength and effective acoustic attenuation requires careful optimization of geometry and material properties. Additionally, large-scale manufacturing and integration into real-world systems require further research and development.

6.4.3 Mechanical–Electromagnetic Metamaterials

Mechanical–electromagnetic metamaterials are a class of multifunctional metamaterials designed to simultaneously control mechanical properties and electromagnetic wave interactions through carefully engineered structural architectures. These metamaterials combine mechanical functionality such as stiffness, flexibility, or load-bearing capacity with the ability to manipulate electromagnetic waves including radio waves, microwaves, and optical frequencies. This coupling between mechanical and electromagnetic behavior enables the development of advanced material systems for applications in communication technologies, sensing systems, aerospace engineering, and smart structures.

In conventional materials, electromagnetic properties are mainly determined by material composition and intrinsic electrical properties such as permittivity and permeability. However, in mechanical–electromagnetic metamaterials, electromagnetic behavior is primarily governed by engineered microstructures and periodic geometries. By designing specific structural patterns, engineers can create materials that interact with electromagnetic waves in unusual ways, including wave absorption, reflection, cloaking, or frequency filtering.

The mechanical component of these metamaterials typically consists of lattice frameworks, cellular architectures, or flexible structural elements that provide mechanical stability and structural support. At the same time, conductive materials, resonant structures, or embedded electromagnetic components are integrated into the architecture to influence electromagnetic wave propagation.

One important feature of mechanical electromagnetic metamaterials is the ability to achieve tunable electromagnetic properties through mechanical deformation. When the metamaterial structure is stretched, compressed, or bent, the geometry of the internal electromagnetic resonators changes. This modification alters the electromagnetic response of the material, allowing engineers to dynamically control properties such as resonant frequency, wave transmission, and electromagnetic shielding performance.

These metamaterials are particularly useful in the design of reconfigurable electromagnetic devices. For example, flexible metamaterial structures can be used to create antennas whose operating frequency can be adjusted by mechanically deforming the structure. Similarly, mechanical deformation can alter the electromagnetic absorption characteristics of metamaterial surfaces used in radar absorption or electromagnetic interference shielding.

Another important application area involves adaptive electromagnetic cloaking systems. Metamaterial structures can be engineered to guide electromagnetic waves around an object, effectively making it less detectable to certain sensing systems. By integrating mechanical control mechanisms, these cloaking properties can be adjusted in response to environmental conditions.

Mechanical–electromagnetic metamaterials also play an important role in structural health monitoring systems. Embedded electromagnetic sensors within metamaterial structures can detect changes in mechanical deformation or structural damage. These systems allow structures to continuously monitor their condition and provide early warning signals when structural issues occur.

Advances in additive manufacturing and multi-material fabrication technologies have significantly facilitated the development of mechanical–electromagnetic metamaterials. These technologies allow conductive materials, dielectric components, and structural frameworks to be integrated into a single metamaterial architecture with high precision.

Despite their promising capabilities, the design of mechanical–electromagnetic metamaterials presents several challenges. Achieving optimal performance requires careful balancing of mechanical strength, electromagnetic functionality, and fabrication constraints. Additionally, accurately modeling the interaction between mechanical deformation and electromagnetic wave behavior often requires advanced multiphysics simulations.

6.4.4 Energy Harvesting Metamaterials

Energy harvesting metamaterials are an emerging class of multifunctional metamaterials designed to capture, convert, and utilize ambient energy from the surrounding environment. These materials integrate engineered structural architectures with energy conversion mechanisms to transform various forms of environmental energy such as mechanical vibrations, thermal gradients, acoustic waves, or electromagnetic radiation into usable electrical energy. This capability enables the development of self-powered systems and sustainable energy solutions for advanced engineering applications.

The energy harvesting capability of metamaterials is largely derived from their engineered periodic structures and resonant geometries. These architectures can amplify mechanical vibrations or concentrate energy within specific regions of the structure, improving the efficiency of energy conversion devices embedded within the metamaterial framework. By carefully designing the internal geometry, engineers can optimize how the metamaterial interacts with external energy sources.

One common approach to energy harvesting in metamaterials involves the integration of piezoelectric materials within lattice structures. Piezoelectric materials generate electrical charge when subjected to mechanical stress or vibration. When embedded within metamaterial architectures, these materials can convert mechanical vibrations—such as those caused by structural movement or environmental disturbances—into electrical energy. This mechanism is particularly useful for harvesting energy from ambient vibrations in buildings, bridges, and transportation systems.

Another important mechanism involves electromagnetic energy harvesting, where metamaterials are designed to capture electromagnetic waves such as radio frequency (RF) signals or microwaves. Specialized metamaterial structures can concentrate electromagnetic energy and direct it toward energy conversion components, improving the efficiency of wireless energy harvesting systems.

Acoustic energy harvesting is another promising application. Mechanical acoustic metamaterials can be engineered to trap and amplify sound waves within resonant cavities. By converting these amplified acoustic vibrations into electrical energy through piezoelectric elements, the metamaterial can harvest energy from ambient noise sources such as industrial machinery, transportation systems, or urban environments.

Thermoelectric energy harvesting is also being explored in metamaterial systems. In this approach, metamaterials are designed to manage heat flow and create localized temperature gradients that drive thermoelectric generators. These generators convert thermal energy directly into electrical power, making them useful in systems where waste heat is readily available.

One of the key advantages of energy harvesting metamaterials is their ability to achieve high energy conversion efficiency through structural resonance and wave manipulation. The periodic architecture of metamaterials allows them to interact with specific frequencies of vibrations or waves, enabling efficient energy concentration and conversion.

Applications of energy harvesting metamaterials include self-powered sensors, wireless monitoring systems, structural health monitoring devices, wearable electronics, and smart infrastructure systems. For example, metamaterial-based energy harvesting devices embedded in bridges or buildings can power monitoring sensors without requiring external power sources.

Advances in additive manufacturing and microfabrication technologies have significantly accelerated the development of energy harvesting metamaterials. These fabrication techniques allow the integration of energy conversion components directly within complex metamaterial structures.

Despite their promising potential, challenges remain in improving energy conversion efficiency, durability under long-term environmental exposure, and large-scale manufacturing feasibility. Researchers continue to explore new metamaterial architectures and hybrid energy harvesting systems to overcome these limitations.

Table 6.3 presents a comparative overview of different multifunctional metamaterial systems used in mechanical metamaterial design. These systems integrate

TABLE 6.3
Comparative Study of Multifunctional Metamaterial Systems for Mechanical Metamaterials

Metamaterial System	Coupled Physical Domains	Working Principle	Key Characteristics	Advantages	Example Applications
Mechanical–Thermal Metamaterials	Mechanical behavior and heat transfer	Engineered lattice structures regulate both structural deformation and thermal conductivity.	Tunable stiffness, thermal insulation, heat flow control.	Enables simultaneous structural support and thermal management.	Energy-efficient building panels, aerospace thermal protection systems.
Mechanical–Acoustic Metamaterials	Mechanical strength and acoustic wave control	Periodic structures and resonant cavities create acoustic bandgaps and vibration attenuation.	Sound absorption, vibration isolation, structural stability.	Provides noise reduction while maintaining load-bearing capability.	Architectural soundproof panels, vibration damping systems in vehicles.
Mechanical–Electromagnetic Metamaterials	Mechanical deformation and electromagnetic wave interaction	Structural geometry alters electromagnetic resonance and wave propagation.	Tunable electromagnetic response, flexible structures.	Enables reconfigurable antennas and electromagnetic shielding.	Adaptive communication devices, radar absorbing structures.
Energy Harvesting Metamaterials	Mechanical vibration and electrical energy generation	Embedded piezoelectric or electromagnetic components convert ambient energy into electricity.	Energy conversion capability, vibration amplification, multifunctional response.	Enables self-powered systems and energy-efficient monitoring devices.	Structural health monitoring sensors, wearable electronics.
Smart Material Integrated Metamaterials	Mechanical properties and stimuli-responsive behavior	Integration of smart materials such as shape memory alloys or piezoelectric elements within metamaterial architectures.	Adaptive stiffness, shape transformation, self-sensing capability.	Allows real-time structural adaptation and programmable behavior.	Adaptive building façades, soft robotic actuators.
Hybrid Multifunctional Metamaterials	Multiple coupled domains (mechanical, thermal, acoustic, electromagnetic)	Combines several functional mechanisms within a single engineered architecture.	Multifunctionality, lightweight design, high efficiency.	Reduces system complexity and improves overall performance.	Advanced aerospace structures, multifunctional architectural systems.

multiple physical domains such as mechanical, thermal, acoustic, electromagnetic, and energy harvesting functions within a single engineered architecture. Such multifunctional designs enable the development of high-performance materials capable of performing multiple roles simultaneously, making them highly valuable for applications in architecture, aerospace engineering, smart infrastructure, robotics, and advanced energy systems.

6.5 ADVANCED MANUFACTURING TECHNOLOGIES FOR METAMATERIALS

The development of mechanical metamaterials relies heavily on advanced manufacturing technologies capable of fabricating complex internal geometries and architected lattice structures. Since the unique properties of metamaterials arise primarily from their engineered microstructures rather than material composition, precise fabrication techniques are essential to achieve the desired structural performance. Traditional manufacturing methods often struggle to produce the intricate and highly periodic architectures required for metamaterial designs. As a result, modern fabrication technologies such as additive manufacturing, microfabrication, and digital fabrication techniques have become critical for the practical realization of metamaterial structures.

One of the most widely used manufacturing approaches for metamaterials is additive manufacturing (AM), commonly known as 3D printing. AM builds structures layer by layer based on digital models, allowing for the fabrication of highly complex geometries that would be difficult or impossible to produce using conventional subtractive methods. Techniques such as selective laser melting (SLM), fused deposition modeling (FDM), stereolithography (SLA), and electron beam melting (EBM) enable the creation of intricate lattice architectures with high precision.

Another important manufacturing method is microfabrication, which is commonly used for producing metamaterials at small scales, particularly in applications involving microelectromechanical systems (MEMS) and microscale metamaterial structures. Techniques such as photolithography, etching, and thin-film deposition allow engineers to create highly detailed microstructures capable of manipulating mechanical, acoustic, or electromagnetic waves.

Laser-based fabrication techniques are also widely employed in metamaterial manufacturing. Laser cutting, laser sintering, and laser direct writing enable the precise fabrication of micro- and mesoscale metamaterial architectures. These methods are particularly useful for producing lattice structures in metals and polymers with high geometric accuracy.

In addition to additive and laser-based methods, robotic and automated manufacturing systems are increasingly used to fabricate large-scale metamaterial structures. Robotic fabrication technologies can construct complex lattice frameworks and modular metamaterial panels for architectural and structural applications. These systems allow for improved scalability and enable the production of metamaterial structures at building or infrastructure scales.

Another emerging manufacturing approach is multi-material fabrication, where different materials with distinct physical properties are combined within a single metamaterial architecture. This technique allows engineers to integrate materials with varying mechanical, thermal, electrical, or responsive properties into a unified structure. Multi-material AM is particularly important for developing multifunctional metamaterials and smart material systems.

Self-assembly techniques are also being explored as a method for producing metamaterials. In this approach, small building blocks or particles spontaneously organize into ordered structures through physical or chemical interactions. Self-assembly processes have the potential to enable large-scale production of metamaterial architectures at micro- and nanoscale levels.

Despite these advancements, several challenges remain in the manufacturing of metamaterials. Fabrication processes must maintain high precision and structural integrity, especially when producing complex geometries at multiple scales. Additionally, manufacturing costs, material limitations, and scalability issues must be addressed to enable widespread industrial adoption.

Ongoing research in hybrid manufacturing technologies, digital fabrication platforms, and advanced material processing techniques continues to expand the possibilities for metamaterial production. As these technologies evolve, they will play a crucial role in enabling the large-scale implementation of metamaterials across various fields including architecture, aerospace engineering, robotics, biomedical devices, and energy systems.

6.5.1 Large-Scale Additive Manufacturing

Large-scale additive manufacturing (LSAM) has emerged as a powerful technology for fabricating mechanical metamaterials with complex lattice structures at structural and architectural scales. AM, commonly referred to as 3D printing, enables the layer-by-layer fabrication of components directly from digital design models. This capability is particularly valuable for metamaterials, whose unique properties rely on intricate internal geometries and periodic architectures that are difficult to produce using conventional manufacturing techniques.

In LSAM, the fabrication process is extended beyond small components to produce large structural elements such as panels, beams, lattice frameworks, and building components. These technologies allow engineers to translate computationally optimized metamaterial designs into full-scale physical structures suitable for applications in architecture, aerospace, automotive engineering, and civil infrastructure.

One of the key advantages of LSAM is its ability to fabricate highly complex geometries without the need for molds or extensive tooling. Since metamaterial performance is determined by geometric configuration rather than bulk material composition, AM provides the flexibility needed to produce periodic lattice structures, cellular architectures, and hierarchical metamaterial designs with high precision.

Several AM techniques are used for large-scale metamaterial fabrication. FDM is commonly used for printing large polymer-based structures by depositing molten thermoplastic materials layer by layer. Selective laser sintering (SLS) and SLM are laser-based techniques that enable the fabrication of metallic metamaterial structures by selectively fusing powdered materials. EBM is another metal-based AM method that uses an electron beam to melt metal powders and build complex components.

LSAM also supports the production of lightweight lattice structures, which are essential for many metamaterial applications. These structures provide high strength-to-weight ratios and efficient load distribution while minimizing material usage. Such properties are particularly beneficial in industries where weight reduction and structural efficiency are critical.

Another important advantage of LSAM is the ability to implement topology-optimized and generatively designed metamaterial architectures. Computational design tools can generate highly optimized lattice structures that maximize performance while minimizing material consumption. AM allows these complex geometries to be fabricated directly from digital design files.

LSAM is also enabling the development of multi-material metamaterials, where different materials with distinct properties are combined within a single structure. This capability allows the integration of structural materials with functional components such as smart materials, sensors, or energy harvesting elements, creating multifunctional metamaterial systems.

Applications of LSAM in metamaterials include architectural façade systems, lightweight structural panels, aerospace components, vibration damping structures, and advanced infrastructure systems. In architectural engineering, LSAM enables the fabrication of customized metamaterial building components that improve structural performance and energy efficiency.

Despite its advantages, LSAM faces several challenges. Achieving high fabrication accuracy, material consistency, and structural integrity over large dimensions can be difficult. Additionally, printing large metamaterial structures requires careful control of thermal stresses, build orientation, and layer bonding quality to ensure reliable performance.

Ongoing advancements in high-speed AM systems, robotic 3D printing technologies, and advanced material processing methods are helping to overcome these challenges. These developments are expanding the capabilities of LSAM and enabling the fabrication of increasingly complex and large-scale metamaterial structures.

6.5.2 Multimaterial and Hybrid Fabrication Techniques

Multi-material and hybrid fabrication techniques represent an important advancement in the manufacturing of mechanical metamaterials, enabling the integration of multiple materials with different physical and functional properties within a single architected structure. Since metamaterials often require combined mechanical, thermal, electrical, or responsive functionalities, the ability to fabricate structures composed of more than one material significantly expands their design possibilities and performance capabilities.

Multi-material fabrication involves the simultaneous or sequential deposition of different materials during the manufacturing process. Each material can be selected to provide specific properties such as structural strength, flexibility, thermal conductivity, electrical conductivity, or responsiveness to environmental stimuli. When these materials are combined within carefully designed metamaterial architectures, the resulting system can exhibit multifunctional behavior that cannot be achieved using a single material alone.

One of the most widely used methods for multi-material fabrication is multi-material AM. Advanced 3D printing systems can deposit different materials in controlled patterns during the printing process. For example, rigid polymers may be combined with flexible elastomers to create metamaterials that exhibit both structural support and controlled deformation capabilities. Similarly, conductive materials can be integrated with structural components to enable electrical functionality and sensing capabilities within metamaterial structures.

Another important approach is hybrid fabrication, which combines multiple manufacturing techniques to produce complex metamaterial architectures. In hybrid manufacturing systems, AM may be combined with traditional manufacturing processes such as laser cutting, machining, molding, or robotic assembly. This approach allows engineers to take advantage of the strengths of different fabrication methods while overcoming the limitations of individual processes.

For instance, AM can be used to fabricate intricate lattice frameworks, while conventional machining can be applied to produce high-precision mechanical components that are later integrated into the metamaterial structure. Similarly, laser-based fabrication techniques can be used to create microscale structural features, which are then combined with larger structural elements fabricated using AM.

Multi-material and hybrid fabrication techniques are particularly valuable in the development of smart metamaterials and stimuli-responsive systems. By embedding materials such as shape memory alloys, piezoelectric materials, or electroactive polymers within structural metamaterial frameworks, engineers can create adaptive systems capable of changing their properties or configuration in response to external stimuli.

Another significant advantage of these fabrication techniques is the ability to develop graded metamaterial structures. Functionally graded materials (FGMs) allow the gradual variation of material properties across the structure, enabling improved stress distribution, enhanced thermal performance, and optimized mechanical behavior. Hybrid fabrication methods make it possible to create such graded architectures with high precision.

Applications of multi-material and hybrid fabrication in metamaterials include adaptive architectural panels, soft robotic systems, biomedical devices, aerospace components, and multifunctional structural systems. In architecture, these techniques enable the production of metamaterial building elements that combine structural support, thermal insulation, and responsive environmental control within a single component.

Despite their advantages, several challenges remain in the development of multi-material and hybrid metamaterial fabrication systems. Issues related to material compatibility, bonding strength between different materials, fabrication accuracy, and manufacturing cost must be carefully addressed to ensure reliable performance.

Ongoing research in advanced AM systems, robotic fabrication platforms, and digital material design continues to expand the capabilities of multi-material and hybrid fabrication techniques. As these technologies evolve, they are expected to play a crucial role in enabling the production of highly complex, multifunctional metamaterial systems for next-generation engineering and architectural applications.

6.5.3 Nanoscale and Microfabrication Approaches

Nanoscale and microfabrication approaches are critical for the development of metamaterials with extremely small structural features, where the internal architecture operates at micro- or nanometer scales. At these scales, metamaterial properties can be engineered to manipulate mechanical, thermal, acoustic, and electromagnetic responses with high precision. The ability to fabricate such fine structures enables the creation of metamaterials with unique and highly tunable properties that are not achievable through conventional manufacturing techniques.

Microfabrication techniques are widely used in the production of microelectromechanical systems (MEMS) and microscale metamaterial structures. These processes involve the precise patterning and structuring of materials using advanced methods such as photolithography, chemical etching, and thin-film deposition. Through these techniques, engineers can create periodic microstructures that influence the propagation of waves, mechanical deformation behavior, and energy transfer mechanisms within the metamaterial.

One of the most common microfabrication methods is photolithography, which uses light to transfer geometric patterns from a photomask onto a photosensitive material layer. This method allows the fabrication of highly precise microstructures that form the building blocks of metamaterial architectures. Photolithography is widely used in the production of microscale lattice structures, microresonators, and functional metamaterial components.

Another important microfabrication technique is etching, which removes material selectively to create desired structural features. Both wet etching and dry etching methods are commonly used to shape microstructures in silicon, polymers, or metal films. Etching processes enable the creation of complex microscale geometries that influence the mechanical and physical properties of metamaterials.

At even smaller scales, nanofabrication techniques are used to produce metamaterials with structural features on the nanometer scale. These techniques include electron beam lithography, nanoimprint lithography, and focused ion beam machining. Nanofabrication enables the creation of extremely fine patterns capable of manipulating physical phenomena such as optical waves, electromagnetic radiation, and nanoscale mechanical behavior.

Self-assembly techniques are also being explored as a method for producing nanoscale metamaterials. In self-assembly processes, small particles or molecular structures organize themselves into ordered patterns through physical or chemical interactions. This approach allows the formation of complex nanoscale metamaterial architectures without the need for extensive external fabrication processes.

Micro- and nanoscale fabrication techniques also support the integration of functional materials and responsive elements within metamaterial structures. For example, microfabricated metamaterials may incorporate piezoelectric layers, conductive coatings, or nanoscale sensors, enabling the development of smart metamaterials with sensing and actuation capabilities.

Applications of nanoscale and microfabricated metamaterials include microsensors, biomedical devices, optical metamaterials, microrobotic systems, and advanced electronic components. In biomedical engineering, microfabricated metamaterials can be used to create devices for drug delivery, tissue engineering, and medical diagnostics. In electronics and photonics, nanoscale metamaterials are used to control electromagnetic waves and enhance device performance.

Despite the significant advantages of nanoscale and microfabrication techniques, several challenges remain. Fabrication at extremely small scales requires high precision, specialized equipment, and strict environmental control, which can increase production complexity and cost. Additionally, scaling these fabrication methods for large-scale manufacturing remains a significant challenge.

Ongoing advancements in nanotechnology, microfabrication tools, and hybrid manufacturing techniques continue to expand the capabilities of metamaterial fabrication at small scales. As these technologies evolve, nanoscale and microfabrication approaches will play a crucial role in enabling the development of next-generation metamaterials with highly precise and multifunctional properties for advanced engineering applications.

Table 6.4 presents a comparative overview of major advanced manufacturing technologies used in the fabrication of mechanical metamaterials. These techniques enable the production of complex architected structures at different length scales, ranging from nanoscale components to large structural systems. The integration of AM, microfabrication, and hybrid fabrication methods plays a crucial role in translating computational metamaterial designs into practical engineering applications in architecture, aerospace, robotics, and advanced structural systems.

6.6 RELIABILITY AND DURABILITY OF MECHANICAL METAMATERIALS

Reliability and durability are critical factors in the practical implementation of mechanical metamaterials, especially when these materials are used in structural, architectural, and engineering applications. While metamaterials exhibit unique mechanical properties due to their engineered internal geometries and periodic lattice structures, their long-term performance must be carefully evaluated to ensure consistent functionality under real-world operating conditions.

Reliability refers to the ability of a metamaterial structure to perform its intended function without failure over a specified period, while durability relates to the material's capacity to withstand repeated mechanical loads, environmental exposure, and long-term operational stresses. Because metamaterials often contain complex architectures composed of thin struts, nodes, and cellular frameworks, understanding their reliability requires detailed analysis of both material properties and structural geometry.

One of the primary concerns in the durability of mechanical metamaterials is fatigue behavior. Many metamaterial structures experience repeated loading cycles during operation, such as vibration, compression, or bending. Over time, these cyclic loads can lead to the initiation and propagation of microcracks within structural elements, potentially causing structural degradation or failure. Researchers often use fatigue testing and computational simulations to evaluate how metamaterial architectures respond to repeated loading conditions.

TABLE 6.4
Comparative Study of Advanced Manufacturing Technologies for Mechanical Metamaterials

Manufacturing Technology	Working Principle	Key Features	Advantages	Limitations	Typical Applications in Mechanical Metamaterials
Additive manufacturing (3D printing)	Layer-by-layer fabrication of structures directly from digital design models.	Supports complex lattice geometries and customized metamaterial architectures.	High design flexibility, minimal material waste, rapid prototyping.	Limited build speed and sometimes lower material strength compared to traditional methods.	Fabrication of lattice-based mechanical metamaterials and lightweight structural components.
Large-scale additive manufacturing (LSAM)	Extension of additive manufacturing for producing large structural components and metamaterial frameworks.	Capable of producing large panels, beams, and architectural structures.	Enables fabrication of large-scale metamaterial systems with complex geometry.	Requires advanced equipment and precise control of thermal stresses.	Architectural metamaterial panels, aerospace structural components.
Multimaterial additive manufacturing	Simultaneous deposition of different materials during the fabrication process.	Allows integration of structural, conductive, and responsive materials in a single structure.	Enables multifunctional metamaterials with combined properties.	Material compatibility and bonding challenges between different materials.	Smart metamaterials, adaptive structures, energy harvesting systems.
Hybrid manufacturing techniques	Combination of additive manufacturing with conventional processes such as machining or laser cutting.	Integrates multiple fabrication methods to produce complex architectures.	Improved precision and structural performance.	Increased manufacturing complexity and process coordination.	Multifunctional metamaterial components and complex engineering systems.
Microfabrication techniques	Fabrication of microscale structures using lithography, etching, and thin-film deposition.	Enables precise control of microscale metamaterial architectures.	High accuracy for small-scale structures and MEMS devices.	Limited scalability for large structural components.	Microelectromechanical metamaterials, sensors, microscale mechanical systems.
Nanofabrication techniques	Fabrication of nanoscale structures using advanced techniques such as electron beam lithography.	Produces extremely fine structural features at nanometer scales.	Enables control of optical, electromagnetic, and nanoscale mechanical properties.	Expensive equipment and complex fabrication processes.	Optical metamaterials, nanoscale sensors, photonic devices.

Another important factor affecting reliability is structural stability under extreme loading conditions. Metamaterial architectures may undergo large deformations, buckling, or instability when subjected to high mechanical stresses. While some metamaterials are intentionally designed to exploit controlled buckling mechanisms, uncontrolled instability can compromise structural performance. Therefore, careful design and optimization are required to ensure that deformation behavior remains predictable and stable.

Environmental factors also play a significant role in determining the durability of metamaterials. Structures exposed to temperature fluctuations, humidity, corrosion, ultraviolet radiation, or chemical exposure may experience material degradation over time. For example, polymer-based metamaterials may soften at elevated temperatures, while metal-based metamaterials may suffer from corrosion or oxidation in harsh environments.

Material selection is therefore a critical consideration in improving the durability of metamaterial systems. Engineers often choose materials with high fatigue resistance, corrosion resistance, and thermal stability to ensure reliable performance under challenging operating conditions. In some cases, protective coatings or surface treatments are applied to enhance the environmental resistance of metamaterial structures.

The manufacturing process also influences the reliability of mechanical metamaterials. Imperfections introduced during fabrication such as microvoids, residual stresses, or geometric inaccuracies can significantly affect structural performance. Advanced manufacturing methods, including high-precision additive manufacturing and post-processing techniques, are used to reduce defects and improve structural integrity.

Another emerging area of research focuses on self-healing metamaterials. These materials incorporate mechanisms that allow the structure to repair minor damage automatically, improving long-term durability. Self-healing approaches may involve the use of microcapsules containing healing agents, reversible chemical bonds, or responsive polymer systems that restore structural integrity after damage occurs.

Computational modeling and simulation play an important role in evaluating metamaterial reliability. Techniques such as FEA, fatigue modeling, and probabilistic reliability analysis help predict how metamaterial structures behave under long-term operational conditions. These tools allow engineers to optimize designs to minimize failure risks and improve structural longevity.

6.6.1 Fatigue and Structural Stability

Fatigue and structural stability are important considerations in the long-term performance of mechanical metamaterials, particularly when these materials are used in applications involving repeated mechanical loading, dynamic forces, and environmental stress conditions. Because metamaterials often consist of complex lattice structures, thin struts, and interconnected unit cells, their mechanical behavior under cyclic loading differs significantly from that of conventional bulk materials.

Fatigue refers to the gradual degradation and eventual failure of a material or structure due to repeated cycles of loading and unloading. In mechanical metamaterials, fatigue behavior is strongly influenced by the geometry of the lattice architecture, material properties, and stress distribution within the structural elements. When a metamaterial structure is subjected to repeated stresses, localized stress concentrations may occur at joints, nodes, or thin structural members. Over time, these stress concentrations can initiate microcracks that propagate through the structure, eventually leading to structural failure.

The fatigue performance of metamaterials depends on several factors, including unit-cell geometry, strut thickness, load distribution, and manufacturing quality. For example, lattice architectures with smooth load paths and optimized geometries tend to distribute stresses more evenly, improving fatigue resistance. In contrast, structures with sharp geometric features or uneven load distribution may experience premature fatigue failure.

Structural stability is another critical aspect of metamaterial performance. Structural stability refers to the ability of a metamaterial structure to maintain its shape and mechanical integrity under external loads. Because many metamaterials rely on lightweight cellular architectures, they may be susceptible to buckling or instability under compressive forces. Buckling occurs when structural elements suddenly deform under compressive loading, potentially leading to loss of load-bearing capacity.

In some metamaterial designs, controlled buckling mechanisms are intentionally incorporated to achieve specific mechanical behaviors, such as energy absorption or shape morphing. However, uncontrolled or unexpected buckling can compromise the stability of the structure. Therefore, engineers must carefully design metamaterial architectures to ensure that buckling behavior remains predictable and within safe operating limits.

To evaluate fatigue and structural stability, researchers often use experimental testing and computational simulations. Fatigue tests subject metamaterial samples to repeated loading cycles to measure their durability and identify failure mechanisms. Meanwhile, numerical methods such as FEA are used to simulate stress distribution, deformation patterns, and buckling behavior within complex metamaterial architectures.

Material selection also plays a crucial role in improving fatigue resistance and structural stability. Materials with high strength, good ductility, and resistance to crack propagation are often preferred for metamaterial structures subjected to cyclic loading. In addition, surface treatments and post-processing techniques may be applied to reduce stress concentrations and improve structural reliability.

Advances in topology optimization and computational design have further improved the ability to design metamaterial structures with enhanced fatigue performance and stability. By optimizing the internal geometry of metamaterial unit cells, engineers can reduce stress concentrations and improve load distribution throughout the structure.

6.6.2 Environmental Effects and Material Degradation

Environmental conditions play a significant role in influencing the long-term reliability and durability of mechanical metamaterials. Since metamaterials often consist of intricate lattice structures, thin struts, and interconnected nodes, their structural integrity can be more sensitive to environmental exposure compared to conventional bulk materials. Environmental effects such as temperature variations, humidity, corrosion, ultraviolet radiation, and chemical exposure can gradually degrade the mechanical performance of metamaterial structures.

One of the most important environmental factors affecting metamaterials is temperature variation. Changes in temperature can cause thermal expansion or contraction of structural components, which may lead to thermal stresses within the metamaterial architecture. In extreme conditions, repeated thermal cycling can induce fatigue and microstructural damage. Polymer-based metamaterials may soften or lose stiffness at elevated temperatures, while metal-based metamaterials may experience thermal fatigue under fluctuating temperature conditions.

Moisture and humidity can also significantly influence metamaterial performance, particularly for materials made from polymers, composites, or porous structures. Moisture absorption may lead to swelling, material softening, or degradation of mechanical properties. In some cases, trapped moisture within cellular structures can promote corrosion or microbial growth, which further compromises structural stability.

Another important environmental concern is corrosion, especially in metallic metamaterials used in structural or outdoor applications. Exposure to oxygen, moisture, or chemical agents can cause corrosion of metallic components, leading to material weakening and loss of load-bearing capacity. Because metamaterials often contain thin structural elements with high surface area-to-volume ratios, they may be more susceptible to corrosion effects compared to solid materials.

Ultraviolet (UV) radiation is another factor that can degrade certain metamaterial materials, particularly polymers and composite materials. Prolonged exposure to UV radiation can cause chemical degradation, discoloration, and reduction in mechanical strength. This is especially important for metamaterials used in outdoor environments such as architectural structures or aerospace applications.

Chemical exposure can also contribute to material degradation. In industrial environments, metamaterial structures may be exposed to acids, solvents, or other reactive chemicals that can alter material properties or cause structural damage. Chemical reactions may weaken the bonding between structural elements or cause changes in the microstructure of the material.

To mitigate environmental degradation, engineers often employ protective coatings, corrosion-resistant materials, and surface treatments. For example, metallic metamaterials may be coated with protective layers to prevent oxidation or corrosion, while polymer-based structures may incorporate UV-resistant additives to improve durability.

Advanced material selection and design optimization are also important strategies for improving environmental resistance. Materials with high chemical stability, corrosion resistance, and thermal durability are preferred for applications exposed to harsh environments.

In addition, computational modeling and environmental testing are commonly used to evaluate the long-term behavior of metamaterials. Accelerated aging tests simulate environmental exposure conditions to assess how metamaterial structures perform over extended periods.

6.6.3 Structural Health Monitoring of Metamaterial Systems

Structural health monitoring (SHM) is an important technique used to evaluate the condition, performance, and safety of engineering structures over time. In the context of mechanical metamaterials, SHM plays a critical role in ensuring that complex lattice architectures and architected material systems maintain their functionality and structural integrity throughout their operational lifespan. Because metamaterials often contain intricate geometries and thin structural elements, early detection of damage or degradation is essential to prevent failure and maintain reliability.

SHM systems are designed to continuously or periodically assess the condition of metamaterial structures using embedded sensors, data acquisition systems, and analytical algorithms. These systems detect changes in structural behavior that may indicate damage, fatigue, or material degradation. In metamaterial structures, monitoring techniques focus on identifying changes in stress distribution, vibration patterns, deformation behavior, or wave propagation characteristics.

One common approach to SHM in metamaterial systems involves the use of embedded sensors integrated directly into the metamaterial architecture. Sensors such as strain gauges, piezoelectric sensors, fiber optic sensors, and microelectromechanical sensors can be incorporated into the structure during fabrication. These sensors measure mechanical responses such as strain, displacement, or vibration, providing real-time information about the structural condition.

Piezoelectric sensors are particularly useful in metamaterial health monitoring because they can both detect and generate mechanical waves. These sensors are capable of monitoring changes in vibration signals or wave propagation patterns within the metamaterial structure. Any deviation from the expected signal pattern may indicate the presence of cracks, material fatigue, or structural defects.

Another advanced SHM technique involves the use of acoustic emission monitoring, where sensors detect high-frequency stress waves generated by crack formation or structural damage. By analyzing these acoustic signals, engineers can identify the location and severity of damage within the metamaterial structure.

Optical sensing technologies, such as fiber optic sensors and laser-based measurement systems, are also widely used for SHM. Fiber optic sensors can measure strain, temperature, and deformation along different sections of a metamaterial structure with high precision. These sensors are particularly useful for monitoring large-scale metamaterial structures used in architecture or infrastructure.

Data collected from SHM systems are often processed using advanced signal analysis and ML algorithms. These analytical tools help identify patterns associated with structural damage, enabling predictive maintenance strategies. ML techniques can also analyze large datasets to predict potential failure mechanisms before they occur.

The integration of SHM systems with metamaterial architectures enables the development of self-monitoring and intelligent structures. In such systems, sensors continuously collect data about structural conditions, and automated control systems may adjust structural responses or trigger maintenance actions when abnormalities are detected.

SHM is particularly important for metamaterials used in critical applications such as aerospace structures, smart buildings, transportation systems, and energy infrastructure. Continuous monitoring helps ensure safety, reduce maintenance costs, and extend the operational lifespan of these advanced materials.

Despite its advantages, implementing SHM systems in metamaterial structures presents several challenges. These include the integration of sensors within complex architectures, data interpretation from large sensor networks, and maintaining sensor reliability under long-term operational conditions.

Ongoing research focuses on developing self-sensing metamaterials and integrated sensing architectures, where the structural material itself acts as a sensing system. Such innovations are expected to significantly enhance the reliability and safety of metamaterial-based engineering systems.

6.7 SUSTAINABILITY AND ECO-FRIENDLY METAMATERIALS

Sustainability has become an important consideration in modern engineering and material science due to increasing concerns about resource depletion, environmental impact, and energy consumption. Mechanical metamaterials offer significant potential for sustainable design because their unique properties arise from engineered structural geometry rather than large quantities of raw materials. By optimizing internal architectures and material distribution, metamaterials can achieve high performance with reduced material usage, making them an attractive option for eco-friendly engineering solutions.

One of the primary contributions of metamaterials to sustainability is material efficiency. Traditional structural materials often rely on solid bulk forms, which may require large amounts of raw material to achieve the desired strength. In contrast, metamaterials utilize cellular and lattice architectures that distribute loads efficiently while minimizing material usage. These lightweight structures can provide comparable mechanical strength while significantly reducing resource consumption.

Another important sustainability benefit of metamaterials is energy efficiency. Metamaterial structures can be designed to regulate heat flow, control acoustic waves, and manage mechanical vibrations. In architectural applications, metamaterial-based building components can improve thermal insulation and passive temperature control, reducing the energy required for heating and cooling systems. This contributes to lower energy consumption and reduced greenhouse gas emissions.

Metamaterials also support the development of multifunctional systems, where a single material architecture performs multiple roles such as structural support, thermal insulation, vibration damping, and acoustic control. By combining multiple functions within a single structure, metamaterials reduce the need for additional materials and components, leading to simpler and more resource-efficient designs.

The use of recyclable and biodegradable materials in metamaterial fabrication is another emerging trend in sustainable material development. Researchers are exploring environmentally friendly materials such as biopolymers, natural fiber composites, and recycled polymers for the production of metamaterial structures. These materials help reduce environmental impact while maintaining desirable mechanical properties.

Advanced manufacturing technologies also contribute to sustainability in metamaterial production. AM and digital fabrication techniques allow precise material placement and minimize waste during fabrication. Unlike traditional subtractive manufacturing methods that remove excess material, AM builds structures layer by layer, significantly reducing raw material waste.

Metamaterials can also support renewable energy technologies and energy harvesting systems. For example, metamaterial architectures can be designed to capture mechanical vibrations or acoustic energy and convert them into electrical energy through embedded piezoelectric elements. Such systems contribute to the development of self-powered devices and sustainable energy solutions.

Another important sustainability advantage of metamaterials is their ability to improve structural longevity and durability. Optimized metamaterial architectures can distribute loads more effectively and reduce stress concentrations, which helps extend the lifespan of structural components. Longer-lasting structures reduce the need for frequent replacement and maintenance, further contributing to resource conservation.

Despite these advantages, challenges remain in the development of eco-friendly metamaterials. Issues related to material availability, large-scale manufacturing, recyclability, and lifecycle assessment must be carefully addressed to ensure that metamaterial technologies truly support sustainable development.

Ongoing research in green materials, circular manufacturing systems, and sustainable design methodologies is helping overcome these challenges. By integrating environmentally friendly materials with optimized metamaterial architectures, engineers can develop high-performance materials that support sustainable engineering practices.

6.7.1 Lightweight Design for Energy Efficiency

Lightweight design is one of the most significant advantages offered by mechanical metamaterials, particularly in applications where reducing structural weight directly contributes to improved energy efficiency and resource conservation. Unlike conventional materials that rely on bulk mass for strength, metamaterials utilize engineered internal geometries and lattice architectures to achieve high strength-to-weight ratios while minimizing material usage.

The lightweight nature of metamaterials arises from their cellular and periodic structural configurations. These architectures distribute mechanical loads efficiently through interconnected struts and nodes, allowing the structure to maintain mechanical stability while using significantly less material. As a result, metamaterial-based structures can provide comparable or even superior performance compared to traditional solid materials while reducing overall weight.

In many engineering systems, reducing structural weight leads to significant improvements in energy efficiency. For example, in transportation systems such as aircraft, automobiles, and trains, lighter structures require less energy for propulsion. By incorporating lightweight metamaterial components into these systems, engineers can reduce fuel consumption and lower greenhouse gas emissions.

Lightweight metamaterial structures are also beneficial in architectural and building applications. In building construction, reducing the weight of structural components can decrease the load on foundations and supporting elements, enabling more efficient use of construction materials. Additionally, metamaterial-based building panels can provide both structural support and thermal insulation, improving the overall energy performance of buildings.

Another advantage of lightweight metamaterial design is its contribution to material efficiency and sustainability. Because metamaterials rely on optimized structural geometry rather than large quantities of raw material, they significantly reduce the amount of material required to achieve desired mechanical properties. This reduction in material consumption helps conserve natural resources and decreases the environmental impact associated with material extraction and processing.

Lightweight metamaterial structures also play an important role in energy-efficient mechanical systems. For example, rotating machinery and mechanical components benefit from reduced inertial loads when lighter materials are used. Lower inertia leads to improved dynamic performance and reduced energy consumption during operation.

Advances in topology optimization and computational design methods have further enhanced the development of lightweight metamaterial architectures. These techniques allow engineers to identify optimal material distributions within a design space, ensuring that material is placed only where it is needed for structural performance. As a result, highly efficient structures with minimal material usage can be created.

Modern AM technologies have also contributed to the realization of lightweight metamaterial designs. These fabrication methods allow the production of intricate lattice structures that would be difficult or impossible to manufacture using conventional techniques.

Despite these advantages, designing lightweight metamaterials requires careful consideration of factors such as structural stability, fatigue resistance, and manufacturing constraints. Engineers must ensure that weight reduction does not compromise structural integrity or durability.

6.7.2 Recyclable and Bio-Inspired Metamaterials

Recyclable and bio-inspired metamaterials represent an important advancement in the development of sustainable and environmentally friendly mechanical metamaterials. As concerns about environmental impact and resource conservation continue to grow, researchers are exploring new approaches that combine recyclable materials and nature-inspired design principles to create high-performance metamaterial systems with reduced ecological footprints.

Recyclable metamaterials are designed using materials that can be reprocessed, reused, or repurposed after the end of their service life. Traditional structural materials often contribute to waste generation when components are discarded after use. In contrast, recyclable metamaterials use materials such as recyclable polymers, lightweight metals, biodegradable composites, and recycled plastics, allowing the material components to be recovered and reused in future manufacturing processes.

The use of recyclable materials in metamaterial architectures also aligns with the concept of circular manufacturing systems, where materials are continuously cycled through production, use, and recycling stages. This approach helps reduce raw material consumption and minimizes environmental pollution associated with manufacturing and disposal processes.

Another important direction in sustainable metamaterial design involves bio-inspired metamaterials, which draw inspiration from structures found in nature. Natural materials such as bone, wood, honeycomb, and plant tissues often exhibit efficient structural designs that combine lightweight properties with high mechanical strength. By studying these natural architectures, engineers can develop metamaterial structures that mimic their efficient load distribution and hierarchical organization.

For example, honeycomb structures found in beehives have inspired many metamaterial lattice designs due to their excellent strength-to-weight ratios and efficient material usage. Similarly, the internal structure of bone tissue provides inspiration for hierarchical lattice metamaterials that exhibit both strength and energy absorption capabilities.

Bio-inspired design also enables the development of metamaterials with adaptive and multifunctional properties. Many natural structures respond dynamically to environmental conditions, and these mechanisms can be replicated in engineered metamaterials. For instance, plant structures that adjust their shape in response to humidity changes have inspired moisture-responsive metamaterials used in adaptive architectural systems.

In addition to structural efficiency, bio-inspired metamaterials can also enhance energy absorption and impact resistance. Many biological systems are designed to dissipate energy efficiently during impacts or environmental stresses. By replicating these mechanisms in metamaterial architectures, engineers can develop protective structures with improved safety performance.

Advanced manufacturing technologies such as AM and digital fabrication have played a crucial role in enabling the production of bio-inspired metamaterial architectures. These technologies allow the fabrication of complex hierarchical structures that closely resemble natural geometries.

Despite their advantages, several challenges remain in the development of recyclable and bio-inspired metamaterials. Material selection, manufacturing scalability, and maintaining structural performance while using sustainable materials require careful design and optimization.

Ongoing research in green materials, sustainable manufacturing processes, and biomimetic design methodologies continues to expand the possibilities of eco-friendly metamaterials. By integrating recyclable materials with nature-inspired structural architectures, engineers can develop metamaterial systems that combine high performance with environmental sustainability.

Figure 6.1 illustrates the concept of recyclable and bio-inspired metamaterials, highlighting how sustainable material selection and nature-inspired structural designs can contribute to environmentally friendly engineering solutions. The figure demonstrates the integration of recyclable materials, biological design inspiration, and sustainable applications, emphasizing the role of metamaterials in promoting eco-friendly technological development.

The left section of the figure represents recyclable metamaterials, which are developed using materials that can be reused or reprocessed after their service life. Examples shown include recycled polymers, biodegradable composites, and recyclable metals.

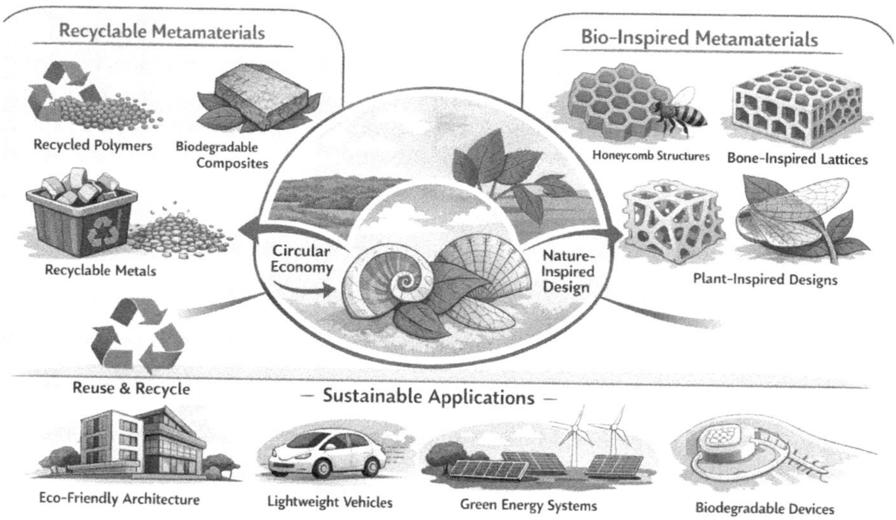

FIGURE 6.1 Recyclable and bio-inspired metamaterials.

These materials can be incorporated into metamaterial architectures such as lattice structures or cellular frameworks while maintaining mechanical strength and structural functionality. The use of recyclable materials supports the concept of a circular economy, where materials are continuously reused rather than discarded, reducing waste and environmental impact.

At the center of the figure, the circular diagram highlights the relationship between circular economy principles and nature-inspired design approaches. The circular layout symbolizes sustainable material cycles in which resources are reused, recycled, and integrated into new metamaterial structures. This concept promotes efficient material utilization and reduces dependence on non-renewable resources.

The right section of the figure illustrates bio-inspired metamaterials, which are designed by mimicking structural patterns found in nature. Examples shown include honeycomb structures, bone-inspired lattice architectures, and plant-inspired geometries. These natural structures exhibit efficient load distribution, lightweight design, and high mechanical strength. By replicating these biological designs, engineers can develop metamaterials with improved mechanical performance, energy absorption, and structural efficiency.

The lower section of the figure highlights various sustainable applications of recyclable and bio-inspired metamaterials. These include eco-friendly architectural structures, lightweight transportation systems, renewable energy technologies, and biodegradable devices. In architecture, metamaterial panels made from recyclable materials can improve structural efficiency and thermal insulation. In transportation, lightweight metamaterial components can reduce energy consumption and fuel usage. Similarly, bio-inspired metamaterials can enhance the performance of renewable energy systems and biodegradable medical devices.

6.7.3 Sustainable Manufacturing Approaches

Sustainable manufacturing approaches focus on developing environmentally responsible production methods that minimize resource consumption, reduce waste, and lower environmental impact during the fabrication of mechanical metamaterials. As the demand for advanced materials continues to grow, it has become increasingly important to adopt manufacturing strategies that support energy efficiency, material conservation, and environmentally friendly production processes.

One of the key aspects of sustainable manufacturing in metamaterials is the use of AM technologies, which significantly reduce material waste compared to traditional subtractive manufacturing methods. In conventional processes such as machining or milling, excess material is removed from a bulk component, often resulting in substantial material loss. In contrast, AM builds structures layer by layer directly from digital design models, ensuring that material is used only where necessary. This approach is particularly beneficial for metamaterials, where complex lattice geometries require precise material placement.

Another important strategy is the use of recyclable and biodegradable materials during the manufacturing process. Materials such as recycled polymers, biodegradable composites, and eco-friendly metals can be used to fabricate metamaterial structures while reducing environmental impact. These materials support the development of circular manufacturing systems, where materials can be reused or recycled after the end of their service life.

Energy efficiency is also a critical factor in sustainable manufacturing. Modern fabrication systems are increasingly designed to optimize energy consumption during production processes. For example, advanced AM machines incorporate energy-efficient heating systems, optimized printing paths, and automated process control to minimize power usage during fabrication.

Another sustainable approach involves the use of digital design and computational optimization techniques. By using advanced computational tools such as topology optimization and generative design, engineers can develop metamaterial structures that use minimal material while maintaining high structural performance. Optimized designs reduce both raw material usage and energy consumption during manufacturing.

Hybrid manufacturing systems also contribute to sustainable production by combining multiple fabrication methods to improve efficiency and reduce waste. For instance, AM may be used to create complex lattice structures, while precision machining is applied only to critical surfaces requiring high accuracy. This combination allows manufacturers to achieve high-quality components with minimal resource consumption.

Sustainable manufacturing also emphasizes reduction of manufacturing waste and emissions. Recycling unused materials, optimizing production workflows, and implementing environmentally friendly processing techniques can significantly reduce waste generation. Additionally, manufacturers are increasingly adopting green manufacturing practices such as low-emission production processes and environmentally safe material handling.

Lifecycle analysis is another important component of sustainable metamaterial manufacturing. Engineers evaluate the environmental impact of materials and manufacturing processes throughout the entire lifecycle of the product from raw material extraction to production, use, and eventual recycling or disposal. This analysis helps identify opportunities to improve sustainability at each stage of the product lifecycle.

Applications of sustainable manufacturing approaches for metamaterials can be found in architecture, transportation systems, renewable energy technologies, and biomedical devices. In architectural applications, sustainable metamaterial components can reduce construction material usage and improve building energy efficiency. Similarly, lightweight metamaterial components in transportation systems can reduce fuel consumption and emissions.

Despite these advancements, challenges remain in achieving fully sustainable metamaterial production. Issues such as scalability, material availability, manufacturing costs, and energy consumption of advanced fabrication technologies must be carefully addressed.

Ongoing research in green materials, eco-friendly fabrication processes, and circular manufacturing systems continues to improve the sustainability of metamaterial production. By integrating environmentally responsible manufacturing techniques with advanced metamaterial design, engineers can develop high-performance materials that support sustainable engineering and environmentally conscious technological development.

Figure 6.2 illustrates the concept of sustainable manufacturing approaches for mechanical metamaterials, highlighting the integration of environmentally friendly materials, energy-efficient fabrication techniques, and circular manufacturing systems. The figure emphasizes how sustainable production strategies can reduce environmental impact while maintaining high performance in metamaterial structures.

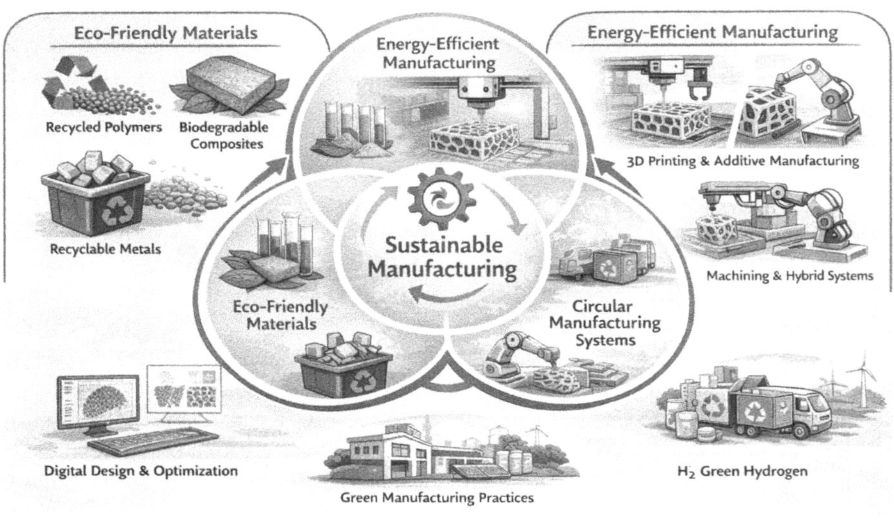

FIGURE 6.2 Sustainable manufacturing approaches for mechanical metamaterials.

The left section of the figure focuses on eco-friendly materials used in sustainable metamaterial manufacturing. These materials include recycled polymers, biodegradable composites, and recyclable metals. Recycled polymers are derived from previously used plastic materials and can be reprocessed into new metamaterial components. Biodegradable composites are designed to decompose naturally over time, reducing environmental pollution. Recyclable metals such as aluminum and steel can be reused after their service life, supporting resource conservation and reducing raw material extraction.

The central portion of the figure represents the concept of energy-efficient manufacturing, which focuses on minimizing energy consumption during production processes. Modern manufacturing technologies such as advanced fabrication systems and automated production lines help optimize energy use while maintaining precision in the fabrication of complex metamaterial architectures. This section emphasizes the role of optimized processing techniques and efficient manufacturing equipment in reducing energy requirements.

The right section of the figure highlights energy-efficient manufacturing technologies, particularly AM (3D printing) and hybrid manufacturing systems. AM enables the layer-by-layer fabrication of complex lattice structures directly from digital models, significantly reducing material waste compared to traditional subtractive processes. Hybrid manufacturing combines AM with conventional machining processes to achieve high precision while maintaining material efficiency.

The lower section of the figure illustrates the role of circular manufacturing systems in sustainable metamaterial production. Circular manufacturing focuses on reusing materials, recycling manufacturing waste, and extending the lifecycle of products. This approach reduces environmental impact by ensuring that materials remain within the production cycle rather than being discarded.

Additional elements shown in the lower part of the figure include digital design and optimization, green manufacturing practices, and renewable energy integration. Digital design tools allow engineers to optimize metamaterial structures to use minimal material while maintaining high structural performance. Green manufacturing practices aim to reduce emissions, waste, and energy consumption during production. Renewable energy sources, such as hydrogen-based energy systems, can further support sustainable industrial operations.

6.8 FUTURE ENGINEERING APPLICATIONS

Mechanical metamaterials are expected to play a transformative role in future engineering systems due to their ability to achieve tailored mechanical properties, multifunctionality, and adaptive performance through engineered architectures. As advances in computational design, smart materials, and advanced manufacturing technologies continue to progress, the range of applications for metamaterials is expanding rapidly across multiple engineering domains.

One of the most promising areas for future applications is aerospace engineering. Lightweight metamaterial structures with optimized lattice geometries can significantly reduce the weight of aircraft and spacecraft components while maintaining high structural strength and stiffness. Such materials can improve fuel efficiency, enhance structural performance, and support the development of deployable aerospace structures such as morphing wings and satellite components.

In architectural and civil engineering, mechanical metamaterials are expected to contribute to the development of smart and resilient infrastructure systems. Metamaterial-based building components can provide improved thermal insulation, vibration damping, and acoustic control. Additionally, seismic metamaterials may be used to design structures capable of mitigating the impact of earthquake waves, enhancing the safety and durability of buildings in earthquake-prone regions.

The field of robotics and soft robotics is another area where metamaterials are expected to have significant impact. Flexible metamaterial architectures can enable the development of robots with programmable motion, adaptive stiffness, and complex deformation capabilities. These features allow robots to interact more safely with humans and operate effectively in dynamic environments.

Mechanical metamaterials are also expected to play a key role in biomedical engineering and healthcare technologies. Metamaterial-based implants, prosthetic devices, and biomedical scaffolds can be designed to mimic the mechanical properties of biological tissues. Such structures may improve the integration of medical implants with surrounding biological tissues and support the development of advanced medical treatments.

Another emerging application involves energy systems and energy harvesting technologies. Metamaterials can be designed to capture mechanical vibrations, acoustic waves, or thermal energy from the environment and convert them into usable electrical energy. These systems may power wireless sensors, wearable electronics, and distributed monitoring devices in smart infrastructure systems.

In transportation engineering, metamaterial-based structures can contribute to the development of lighter and more energy-efficient vehicles. Lightweight lattice structures can reduce vehicle weight, improve fuel efficiency, and enhance crash energy absorption. Metamaterial-based vibration damping systems can also improve passenger comfort by reducing structural vibrations and noise.

Metamaterials are also expected to enable the development of advanced sensing and adaptive systems. By integrating sensors and responsive materials within metamaterial architectures, engineers can create structures capable of self-monitoring and adaptive response to environmental changes. Such intelligent materials could be used in smart infrastructure, aerospace systems, and automated industrial processes.

Advances in AM and digital fabrication technologies will further accelerate the adoption of metamaterials in engineering applications. These technologies enable the fabrication of complex geometries and multifunctional structures that were previously difficult to manufacture using traditional techniques.

Despite these promising opportunities, challenges remain in scaling metamaterial technologies for large-scale industrial applications. Issues related to manufacturing cost, material durability, design complexity, and long-term reliability must be addressed to ensure practical implementation.

6.8.1 Aerospace and Space Structures

Mechanical metamaterials are expected to play a significant role in the development of next-generation aerospace and space structures due to their ability to provide lightweight, high-strength, and multifunctional structural performance. Aerospace systems require materials that can withstand extreme mechanical loads, temperature variations, and harsh environmental conditions while maintaining minimal weight. Metamaterials offer innovative solutions by utilizing engineered lattice architectures and periodic structures that optimize strength-to-weight ratios and structural efficiency.

One of the most important advantages of metamaterials in aerospace applications is weight reduction. Aircraft and spacecraft performance is strongly influenced by structural weight, as lighter components require less energy for propulsion and improve overall fuel efficiency. Metamaterial-based lattice structures can significantly reduce the mass of structural components while maintaining high mechanical stiffness and strength. This lightweight design contributes to improved aerodynamic performance and energy efficiency in aerospace systems.

Another important application of metamaterials in aerospace engineering is the development of deployable and morphing structures. Certain metamaterial architectures can undergo controlled shape transformations in response to mechanical forces or environmental stimuli. These properties enable the design of morphing wings, deployable antennas, and expandable satellite components. Such structures can be compactly stored during launch and later expanded or reconfigured once deployed in space.

Metamaterials also offer significant benefits for vibration and noise control in aerospace systems. Aircraft and spacecraft are often subjected to intense mechanical vibrations caused by engines, aerodynamic forces, and mechanical operations. Mechanical–acoustic metamaterials can be designed to absorb or redirect vibration energy, improving structural stability and reducing noise levels within aerospace structures.

Another critical application area is thermal management in space environments. Spacecraft components are exposed to extreme temperature fluctuations due to direct solar radiation and the vacuum of space. Metamaterial-based thermal management systems can regulate heat transfer through carefully designed architectures, helping maintain stable operating temperatures for sensitive electronic systems.

In addition to thermal control, metamaterials can provide impact protection and energy absorption for aerospace structures. Spacecraft operating in orbit may encounter micrometeoroids or space debris that can cause structural damage. Metamaterial-based protective structures with energy-absorbing lattice geometries can help dissipate impact energy and reduce the risk of catastrophic structural failure.

Metamaterials are also being explored for use in adaptive and smart aerospace structures. By integrating smart materials and sensing technologies into metamaterial architectures, engineers can develop structures capable of self-monitoring and adaptive response. These systems can detect structural damage, monitor stress distribution, and adjust structural properties in response to operational conditions.

Advances in AM technologies have greatly facilitated the development of metamaterial structures for aerospace applications. Large-scale and high-precision AM systems enable the fabrication of complex lattice architectures in lightweight metals and advanced composite materials.

Despite these promising opportunities, several challenges remain in implementing metamaterials in aerospace systems. Issues related to manufacturing scalability, long-term reliability, fatigue resistance, and certification for aerospace standards must be addressed before widespread adoption becomes possible.

6.8.2 Biomedical Devices and Implants

Mechanical metamaterials are increasingly being explored for applications in biomedical devices and implants, where tailored mechanical properties and complex structural architectures can significantly improve medical performance and patient outcomes. In biomedical engineering, materials must often replicate the mechanical behavior of biological tissues, provide biocompatibility, and support long-term functionality within the human body. The unique geometry-driven properties of metamaterials make them well suited for these requirements.

One of the most important applications of metamaterials in biomedical engineering is the development of orthopedic implants, such as bone replacement structures, joint implants, and spinal implants. Conventional implant materials are typically much stiffer than natural bone, which can lead to a phenomenon known as stress shielding. Stress shielding occurs when the implant carries most of the mechanical load, reducing the load experienced by surrounding bone tissue and potentially leading to bone weakening. Metamaterial-based lattice structures can be designed to closely match the elastic modulus of natural bone, thereby improving load transfer and reducing stress shielding effects.

Another significant application involves tissue engineering scaffolds. Tissue engineering aims to regenerate damaged or diseased tissues using artificial scaffolds that support cell growth and tissue formation. Metamaterial scaffolds can be designed with porous and interconnected architectures that allow the flow of nutrients, oxygen, and biological fluids. These structures provide a supportive environment for cell attachment and proliferation, facilitating the regeneration of bone and other tissues.

Mechanical metamaterials also enable the development of customized medical implants. With the help of advanced computational design and AM technologies, patient-specific implant structures can be created based on medical imaging data such as CT scans or MRI scans. These customized metamaterial implants can be optimized to match the anatomical geometry and mechanical requirements of individual patients, improving implant performance and patient comfort.

Another promising area is the development of metamaterial-based prosthetic devices. Prosthetic limbs require materials that are lightweight, durable, and capable of mimicking the mechanical behavior of natural body parts. Metamaterial structures with optimized lattice geometries can provide improved energy absorption, flexibility, and strength, enhancing the performance of prosthetic devices.

Metamaterials are also being investigated for use in medical stents and vascular devices. Stents are small tubular structures used to keep blood vessels open in patients with cardiovascular diseases. Metamaterial-based stents can be designed with flexible and expandable architectures that improve mechanical stability while allowing natural movement of blood vessels.

In addition to structural implants, metamaterials can support the development of smart biomedical devices. By integrating sensing elements and responsive materials within metamaterial architectures, it is possible to create implants capable of monitoring physiological conditions, detecting stress levels, or responding to biological signals. These intelligent medical devices may provide real-time health monitoring and improve treatment outcomes.

The use of biocompatible materials such as titanium alloys, biodegradable polymers, and bioactive ceramics is essential for biomedical metamaterial applications. These materials ensure that implants interact safely with biological tissues and support long-term functionality within the human body.

Despite these promising advancements, several challenges remain in the use of metamaterials for biomedical applications. Issues related to biocompatibility, long-term durability, manufacturing precision, and regulatory approval must be carefully addressed to ensure safe and effective medical use.

6.8.3 Civil Infrastructure and Smart Cities

Mechanical metamaterials are emerging as a promising class of materials for applications in civil infrastructure and smart city development. As modern cities face challenges related to population growth, environmental sustainability, structural safety, and energy efficiency, metamaterials offer innovative solutions through their engineered architectures and multifunctional capabilities. By utilizing carefully designed structural geometries, metamaterials can enhance the performance, durability, and adaptability of infrastructure systems.

One of the key applications of metamaterials in civil infrastructure is the development of lightweight and high-strength structural components. Traditional construction materials such as concrete and steel provide strength but often involve significant material consumption and structural weight. Metamaterial-based lattice structures and cellular architectures can provide comparable mechanical strength while significantly reducing material usage and structural mass. This contributes to efficient material utilization and improved sustainability in construction projects.

Metamaterials also play a significant role in vibration and seismic protection for buildings and infrastructure systems. In earthquake-prone regions, structures must be capable of withstanding seismic waves and minimizing structural damage. Seismic metamaterials are designed to manipulate and redirect seismic wave propagation, potentially reducing the intensity of vibrations transmitted to buildings. These materials can be incorporated into building foundations or surrounding ground structures to improve seismic resilience.

Another important application involves acoustic and vibration control in urban environments. Cities often face problems related to traffic noise, industrial vibrations, and construction activities. Mechanical–acoustic metamaterials can be integrated into building walls, noise barriers, and urban infrastructure systems to absorb and redirect sound waves. This capability can significantly improve acoustic comfort in residential and commercial areas.

Metamaterials also support the development of energy-efficient building systems, which are essential for sustainable urban development. Metamaterial-based building envelopes and façade systems can improve thermal insulation and passive temperature regulation, reducing energy consumption for heating and cooling. These systems contribute to the creation of green buildings and energy-efficient urban infrastructure.

In addition to structural and environmental benefits, metamaterials can enable the development of smart infrastructure systems. By integrating sensors, smart materials, and data communication technologies within metamaterial architectures, engineers can create structures capable of self-monitoring and adaptive response. These smart systems can continuously monitor structural conditions such as stress levels, vibration patterns, and temperature variations.

SHM systems integrated within metamaterial-based infrastructure can detect early signs of damage or material degradation. This capability allows engineers to implement predictive maintenance strategies, improving the safety and lifespan of infrastructure systems while reducing maintenance costs.

Metamaterials may also contribute to the development of adaptive urban systems within smart cities. For example, adaptive façade panels based on metamaterial architectures could regulate sunlight, airflow, and temperature within buildings. Such systems help optimize energy usage and improve indoor environmental quality.

Advances in LSAM and digital construction technologies are further enabling the use of metamaterials in civil infrastructure. These technologies allow the fabrication of complex structural components and modular metamaterial panels for buildings and infrastructure projects.

Despite their potential, several challenges remain in implementing metamaterials in large-scale infrastructure systems. Issues related to manufacturing scalability, long-term durability, cost-effectiveness, and integration with conventional construction methods must be addressed before widespread adoption can occur.

Mechanical metamaterials are playing an increasingly important role in the development of advanced robotics and autonomous systems, where adaptability, lightweight structures, and multifunctional capabilities are essential. Traditional robotic systems are often constructed using rigid materials and mechanical components, which can limit flexibility and adaptability in dynamic environments. In contrast, metamaterials provide engineered structural architectures that enable programmable mechanical behavior, tunable stiffness, and controlled deformation, making them highly suitable for next-generation robotic systems.

One of the key advantages of metamaterials in robotics is their ability to enable lightweight and flexible structural designs. Robots designed for exploration, manufacturing, or service applications benefit significantly from reduced weight and improved mobility. Metamaterial lattice structures can provide high strength-to-weight ratios while maintaining structural stability, allowing robotic systems to operate efficiently while consuming less energy.

Metamaterials are also widely used in the development of soft robotic systems, which are designed to mimic the flexibility and adaptability of biological organisms. Soft robots are constructed from flexible materials that allow them to deform and interact safely with their surroundings. By incorporating metamaterial architectures into soft robotic components, engineers can control deformation patterns and mechanical responses more precisely. This enables robots to perform complex tasks such as grasping delicate objects, navigating irregular environments, and performing medical procedures.

Another important capability offered by metamaterials is tunable stiffness. In many robotic applications, different tasks require different mechanical responses. For example, a robot may need to be flexible when interacting with fragile objects but rigid when lifting heavy loads. Metamaterial structures can be designed to adjust their stiffness dynamically, enabling robots to adapt to changing operational conditions.

Mechanical metamaterials also contribute to the development of adaptive robotic actuators. Actuators are components responsible for generating movement in robotic systems. Metamaterial-based actuators can amplify motion, distribute forces efficiently, and enable complex movement patterns such as bending, twisting, and expansion. These features allow robots to achieve more natural and efficient motion.

In autonomous robotic systems, metamaterials can also enhance vibration control and mechanical stability. Robots operating in industrial environments or rough terrains may experience mechanical vibrations that affect their performance. Metamaterial-based vibration damping systems can reduce these disturbances and improve operational precision.

Another promising application is the integration of metamaterials with sensing technologies. By embedding sensors within metamaterial structures, robotic systems can monitor mechanical stress, deformation, or environmental conditions. This capability enables the development of self-sensing robotic components that provide real-time feedback to control systems.

Metamaterials are also being explored for use in bio-inspired robotic systems. Many natural organisms, such as octopuses and insects, exhibit highly flexible and adaptive movement. By mimicking these biological structures through metamaterial design, engineers can create robots capable of multi-directional movement, shape transformation, and environmental adaptability.

Advances in AM and digital fabrication technologies have significantly accelerated the development of metamaterial-based robotic systems. These technologies allow engineers to fabricate complex metamaterial structures with integrated functional elements, enabling rapid prototyping and customization of robotic components.

Despite their promising capabilities, several challenges remain in implementing metamaterials in robotics and autonomous systems. Issues such as actuation efficiency, control complexity, material durability, and energy supply must be addressed to ensure reliable long-term performance.

6.8.4 Robotics and Autonomous Systems

Mechanical metamaterials are playing an increasingly important role in the development of advanced robotics and autonomous systems, where adaptability, lightweight structures, and multifunctional capabilities are essential. Traditional robotic systems are often constructed using rigid materials and mechanical components, which can limit flexibility and adaptability in dynamic environments. In contrast, metamaterials provide engineered structural architectures that enable programmable mechanical behavior, tunable stiffness, and controlled deformation, making them highly suitable for next-generation robotic systems.

One of the key advantages of metamaterials in robotics is their ability to enable lightweight and flexible structural designs. Robots designed for exploration, manufacturing, or service applications benefit significantly from reduced weight and improved mobility. Metamaterial lattice structures can provide high strength-to-weight ratios while maintaining structural stability, allowing robotic systems to operate efficiently while consuming less energy.

Metamaterials are also widely used in the development of soft robotic systems, which are designed to mimic the flexibility and adaptability of biological organisms. Soft robots are constructed from flexible materials that allow them to deform and interact safely with their surroundings. By incorporating metamaterial architectures into soft robotic components, engineers can control deformation patterns and mechanical responses more precisely. This enables robots to perform complex tasks such as grasping delicate objects, navigating irregular environments, and performing medical procedures.

Another important capability offered by metamaterials is tunable stiffness. In many robotic applications, different tasks require different mechanical responses. For example, a robot may need to be flexible when interacting with fragile objects but rigid when lifting heavy loads. Metamaterial structures can be designed to adjust their stiffness dynamically, enabling robots to adapt to changing operational conditions.

Mechanical metamaterials also contribute to the development of adaptive robotic actuators. Actuators are components responsible for generating movement in robotic systems. Metamaterial-based actuators can amplify motion, distribute forces efficiently, and enable complex movement patterns such as bending, twisting, and expansion. These features allow robots to achieve more natural and efficient motion.

In autonomous robotic systems, metamaterials can also enhance vibration control and mechanical stability. Robots operating in industrial environments or rough terrains may experience mechanical vibrations that affect their performance. Metamaterial-based vibration damping systems can reduce these disturbances and improve operational precision.

Another promising application is the integration of metamaterials with sensing technologies. By embedding sensors within metamaterial structures, robotic systems can monitor mechanical stress, deformation, or environmental conditions. This capability enables the development of self-sensing robotic components that provide real-time feedback to control systems.

Metamaterials are also being explored for use in bio-inspired robotic systems. Many natural organisms, such as octopuses and insects, exhibit highly flexible and adaptive movement. By mimicking these biological structures through metamaterial design, engineers can create robots capable of multi-directional movement, shape transformation, and environmental adaptability.

Advances in AM and digital fabrication technologies have significantly accelerated the development of metamaterial-based robotic systems. These technologies allow engineers to fabricate complex metamaterial structures with integrated functional elements, enabling rapid prototyping and customization of robotic components.

Despite their promising capabilities, several challenges remain in implementing metamaterials in robotics and autonomous systems. Issues such as actuation efficiency, control complexity, material durability, and energy supply must be addressed to ensure reliable long-term performance.

Table 6.5 presents a comparative overview of future engineering applications of mechanical metamaterials across multiple technological domains. The unique properties of metamaterials such as lightweight structures, programmable mechanical behavior, and multifunctionality enable innovative solutions for aerospace engineering, biomedical devices, smart infrastructure, robotics, and energy systems. As advancements in computational design and advanced manufacturing technologies continue, the adoption of metamaterials in these engineering applications is expected to expand significantly.

TABLE 6.5
Future Engineering Applications of Mechanical Metamaterials

Application Area	Key Features of Metamaterials	Engineering Advantages	Typical Applications
Aerospace and Space Structures	Lightweight lattice structures with high strength-to-weight ratios and adaptive configurations.	Reduces structural weight, improves fuel efficiency, and enables deployable structures.	Morphing wings, satellite components, vibration damping panels.
Biomedical Devices and Implants	Porous and customizable lattice structures that mimic biological tissues.	Improves biocompatibility, reduces stress shielding, and supports tissue growth.	Bone implants, prosthetic devices, biomedical scaffolds, vascular stents.
Civil Infrastructure and Smart Cities	Architected structures with vibration control and thermal insulation properties.	Enhances structural safety, energy efficiency, and noise reduction in urban environments.	Earthquake-resistant buildings, acoustic barriers, smart building façades.
Robotics and Autonomous Systems	Flexible and programmable metamaterial architectures with tunable stiffness.	Enables adaptive motion, lightweight robotic components, and safe human–robot interaction.	Soft robots, robotic grippers, adaptive robotic actuators.
Energy Systems and Energy Harvesting	Structures capable of capturing vibration, thermal, or acoustic energy.	Enables self-powered devices and improves energy efficiency.	Energy harvesting sensors, smart monitoring systems, renewable energy devices.
Advanced Transportation Systems	Lightweight and energy-absorbing metamaterial structures.	Improves vehicle efficiency, impact resistance, and vibration control.	Automotive crash structures, lightweight vehicle components, railway vibration damping systems.

6.9 KEY TECHNICAL CHALLENGES AND RESEARCH OPPORTUNITIES

Although mechanical metamaterials have demonstrated remarkable potential across multiple engineering domains, several technical challenges and research opportunities remain before these materials can be widely adopted in large-scale practical applications. Addressing these challenges requires continued advances in computational design, material science, manufacturing technologies, and system integration.

One of the primary challenges in the development of mechanical metamaterials is manufacturing scalability. Many metamaterial structures rely on highly complex lattice geometries and micro-architectures that are difficult to produce using conventional manufacturing techniques. While AM has significantly improved fabrication capabilities, producing metamaterials at large industrial scales while maintaining precision, structural integrity, and cost efficiency remains a major research challenge.

Another important challenge involves material limitations and durability. Metamaterial performance depends not only on structural geometry but also on the mechanical properties of the base materials used in fabrication. Many metamaterial structures are composed of thin structural elements that may be susceptible to fatigue, fracture, and environmental degradation. Developing new materials with improved strength, fatigue resistance, and environmental stability is essential for long-term reliability.

Design complexity is also a significant challenge. The performance of metamaterials is governed by multiple parameters, including unit-cell geometry, lattice arrangement, and material composition. Designing optimized structures often requires advanced computational tools such as topology optimization, generative design algorithms, and machine learning-based design frameworks. Developing more efficient design methodologies that can handle large design spaces and complex performance requirements remains an active area of research.

Another challenge involves the integration of multifunctional properties. Many metamaterial systems are designed to simultaneously provide mechanical, thermal, acoustic, and electromagnetic functionalities. Achieving optimal performance across multiple physical domains without compromising structural integrity is a complex task that requires sophisticated multiphysics modeling and optimization techniques.

Standardization and testing methodologies also present challenges for the practical implementation of metamaterials. Unlike conventional materials, metamaterials often exhibit highly anisotropic and geometry-dependent properties. Developing standardized testing procedures and performance evaluation methods is necessary to ensure reliable characterization and certification for engineering applications.

In addition to these challenges, the cost of fabrication and material processing remains a barrier to large-scale adoption. Advanced manufacturing technologies capable of producing complex metamaterial structures can be expensive, and reducing manufacturing costs is essential for industrial implementation.

Despite these challenges, the field of mechanical metamaterials presents numerous research opportunities. Advances in ML and AI are enabling new approaches for automated metamaterial design and optimization. These technologies can accelerate the discovery of innovative architectures with improved performance characteristics.

Table 6.6 summarizes the major technical challenges and emerging research opportunities in mechanical metamaterials. Addressing these challenges through advances in computational design, advanced manufacturing, smart material integration, and sustainable material development will enable the widespread adoption of metamaterials in future engineering applications such as aerospace structures, smart infrastructure, biomedical devices, and robotics.

TABLE 6.6
Key Technical Challenges and Research Opportunities in Mechanical Metamaterials

Area	Key Technical Challenges	Research Opportunities	Potential Impact
Manufacturing Scalability	Difficulty in producing complex lattice structures at large industrial scales with high precision.	Development of advanced additive manufacturing, robotic fabrication, and automated production systems.	Enables large-scale deployment of metamaterials in infrastructure, aerospace, and industrial applications.
Material Durability and Fatigue Resistance	Thin structural elements may experience fatigue, fracture, and environmental degradation.	Development of high-performance alloys, composites, and self-healing materials.	Improves long-term reliability and structural lifespan of metamaterial systems.
Computational Design Complexity	Large design spaces and complex unit-cell geometries require extensive computational resources.	Application of topology optimization, generative design, and AI-based design algorithms.	Accelerates discovery of optimized metamaterial architectures with improved performance.
Multifunctional Integration	Difficulty in achieving optimal performance across mechanical, thermal, acoustic, and electromagnetic domains simultaneously.	Development of multiphysics modeling and multi-material metamaterial architectures.	Enables multifunctional structures for smart systems and advanced engineering applications.
Standardization and Testing Methods	Lack of standardized testing procedures for geometry-dependent metamaterial properties.	Establishment of international standards and advanced experimental characterization techniques.	Facilitates certification and industrial adoption of metamaterials.
Manufacturing Cost and Economic Feasibility	Advanced fabrication technologies can be expensive and resource-intensive.	Development of cost-effective fabrication methods and scalable manufacturing systems.	Promotes commercial viability and widespread implementation.
Smart and Adaptive Material Integration	Difficulty in integrating sensors, actuators, and responsive materials into metamaterial architectures.	Research on smart metamaterials with embedded sensing and actuation capabilities.	Enables adaptive structures and intelligent engineering systems.
Bio-Inspired and Sustainable Design	Challenges in replicating complex natural architectures and using eco-friendly materials.	Exploration of biomimetic designs and recyclable materials.	Supports sustainable engineering and environmentally friendly material systems.

6.10 OUTLOOK AND FUTURE DIRECTIONS IN MECHANICAL METAMATERIALS

Mechanical metamaterials represent a rapidly evolving area of research that is expected to significantly influence the future of materials engineering, structural design, and advanced technological systems. Over the past decade, advances in computational design, AM, and smart materials have enabled the development of metamaterial architectures with unprecedented mechanical performance and multifunctional capabilities. As research continues to progress, mechanical metamaterials are likely to transition from experimental prototypes to widely adopted engineering materials.

One of the most important future directions in this field is the development of programmable and adaptive metamaterials. These materials can dynamically change their mechanical properties in response to external stimuli such as temperature, magnetic fields, mechanical stress, or electrical signals. Such adaptive capabilities will enable the creation of intelligent structures capable of adjusting their performance in real time, opening new possibilities for applications in aerospace systems, robotics, and smart infrastructure.

Another promising research direction involves the integration of AI and ML in metamaterial design. AI-driven computational models can analyze large datasets of structural configurations and predict optimized metamaterial architectures with specific performance characteristics. This approach can significantly accelerate the discovery of innovative designs and improve the efficiency of the metamaterial development process.

The advancement of multi-material and hybrid metamaterial systems is also expected to play a major role in future developments. By combining materials with different mechanical, thermal, electrical, or responsive properties within a single architecture, engineers can create metamaterials that perform multiple functions simultaneously. These multifunctional systems may be used in smart structures, energy systems, biomedical devices, and advanced robotics.

Another key area of future research is the development of sustainable and environmentally friendly metamaterials. The use of recyclable materials, bio-inspired structural designs, and energy-efficient manufacturing methods will contribute to the creation of metamaterials that support sustainable engineering practices. Such developments align with global efforts to reduce environmental impact and promote circular manufacturing systems.

Advances in LSAM and digital fabrication technologies will further accelerate the adoption of metamaterials in practical applications. These technologies enable the fabrication of complex lattice structures at various scales, from nanoscale components to large structural systems used in architecture and infrastructure projects.

The future of mechanical metamaterials will also involve the development of self-sensing and self-healing structures. By integrating sensing technologies and responsive materials within metamaterial architectures, engineers can create systems capable of monitoring their own structural health and automatically repairing minor damage. These capabilities will significantly improve the reliability and lifespan of critical engineering systems.

In addition, metamaterials are expected to play an important role in next-generation energy systems. Structures designed to harvest mechanical vibrations, acoustic waves, or thermal energy from the environment may provide sustainable energy sources for distributed sensors and smart infrastructure.

Despite these promising developments, several challenges must be addressed to fully realize the potential of mechanical metamaterials. These include improving manufacturing scalability, cost-effectiveness, material durability, and standardization of testing methods. Addressing these challenges will require interdisciplinary collaboration among researchers in materials science, mechanical engineering, computational modeling, and manufacturing technologies.

Table 6.7 summarizes the future research directions and emerging technological developments in mechanical metamaterials. Advances in AI, smart materials, sustainable manufacturing, and large-scale fabrication technologies are expected to drive the next generation of metamaterial systems. These developments will expand the use of metamaterials in aerospace engineering, smart infrastructure, biomedical technologies, robotics, and energy systems, enabling innovative solutions for future engineering challenges.

TABLE 6.7
Outlook and Future Directions in Mechanical Metamaterials

Future Direction	Key Focus Area	Technological Development	Expected Impact
Programmable and Adaptive Metamaterials	Development of structures capable of changing properties in response to external stimuli.	Integration of smart materials such as shape memory alloys and piezoelectric components.	Enables adaptive structures for aerospace, robotics, and smart infrastructure.
AI and Machine Learning–Driven Design	Use of artificial intelligence to optimize metamaterial architectures.	Machine learning algorithms for automated design and property prediction.	Accelerates discovery of high-performance metamaterial structures.
Multi-Material and Hybrid Metamaterials	Integration of multiple materials within a single metamaterial architecture.	Multi-material additive manufacturing and hybrid fabrication techniques.	Enables multifunctional materials with combined mechanical, thermal, and electrical properties.
Sustainable and Eco-Friendly Metamaterials	Development of recyclable, biodegradable, and environmentally friendly materials.	Use of green manufacturing processes and circular material systems.	Supports sustainable engineering and reduces environmental impact.
Large-Scale Manufacturing Technologies	Scaling metamaterial fabrication for industrial applications.	Large-scale additive manufacturing and automated fabrication systems.	Facilitates practical deployment in construction, aerospace, and transportation.

(Continued)

TABLE 6.7 (Continued)
Outlook and Future Directions in Mechanical Metamaterial

Future Direction	Key Focus Area	Technological Development	Expected Impact
Self-Sensing and Self-Healing Metamaterials	Integration of sensing technologies and damage-repair mechanisms.	Embedded sensors, smart materials, and self-healing polymers.	Enhances reliability, durability, and structural health monitoring capabilities.
Energy Harvesting and Smart Energy Systems	Development of metamaterials capable of capturing ambient energy.	Integration of piezoelectric, thermoelectric, and acoustic energy harvesting systems.	Enables self-powered sensors and smart infrastructure systems.
Bio-Inspired and Hierarchical Metamaterials	Adoption of structural concepts inspired by natural materials.	Hierarchical lattice structures based on bone, plant, and honeycomb architectures.	Improves strength-to-weight ratio and structural efficiency.

6.11 SUMMARY

This chapter examined the emerging trends, technological advancements, and key challenges associated with the future development of mechanical metamaterials. Mechanical metamaterials are unique engineered materials whose properties are primarily governed by carefully designed internal architectures rather than conventional material composition. As research in this field continues to expand, these materials are expected to play an important role in the development of advanced engineering systems, smart structures, and sustainable technologies.

The chapter first discussed the role of advanced computational design and optimization techniques in the development of next-generation metamaterials. Methods such as topology optimization, generative design algorithms, FEM, and machine learning-based design approaches enable engineers to explore large design spaces and identify optimized metamaterial architectures with tailored mechanical properties. These computational tools significantly accelerate the discovery and development of high-performance metamaterial structures.

Another major focus of the chapter was the development of programmable and adaptive metamaterials. These materials incorporate smart materials and responsive mechanisms that allow them to change their mechanical properties, shape, or stiffness in response to external stimuli such as temperature, electrical signals, or mechanical forces. Such capabilities open new opportunities for applications in soft robotics, adaptive structures, aerospace systems, and smart infrastructure.

The chapter also explored the concept of multifunctional metamaterial systems, where a single material architecture can simultaneously provide multiple functionalities such as mechanical strength, thermal regulation, acoustic control, electromagnetic response, and energy harvesting. The integration of these capabilities within a single structure reduces system complexity and improves overall efficiency.

Advancements in advanced manufacturing technologies were also discussed as a key enabler for metamaterial development. Fabrication methods such as LSAM, multi-material fabrication, hybrid manufacturing systems, and nanoscale microfabrication techniques allow the production of complex architected materials across multiple length scales. These technologies are critical for translating computational metamaterial designs into practical engineering components.

The chapter further addressed important considerations related to the reliability and durability of metamaterial structures, including fatigue behavior, structural stability, environmental effects, and SHM systems. Understanding these aspects is essential to ensure long-term performance and safety when metamaterials are used in real-world engineering applications.

Sustainability was another important theme discussed in the chapter. The development of lightweight metamaterial architectures, recyclable materials, bio-inspired designs, and sustainable manufacturing approaches contributes to resource efficiency and reduced environmental impact. These features align with global efforts to promote sustainable engineering and environmentally responsible technologies.

In addition, the chapter explored several future engineering applications of mechanical metamaterials across diverse fields including aerospace structures, biomedical implants, civil infrastructure, robotics, and energy systems. These applications highlight the transformative potential of metamaterials in improving structural performance, energy efficiency, and system adaptability.

Finally, the chapter examined the key technical challenges and future research opportunities in this field. Challenges such as manufacturing scalability, material durability, design complexity, and standardization must be addressed to enable widespread adoption. At the same time, emerging research areas including AI-driven design, smart and adaptive metamaterials, self-healing materials, and sustainable metamaterial systems offer exciting opportunities for future innovation.

Overall, the future of mechanical metamaterials is highly promising. Continued advancements in computational design, advanced manufacturing, smart material integration, and sustainable engineering practices are expected to drive the development of next-generation metamaterial systems that will significantly impact engineering, architecture, robotics, healthcare, and energy technologies.

REVIEW QUESTIONS

1 What are mechanical metamaterials and why are they considered important for future engineering applications?
2 How do mechanical metamaterials differ from conventional engineering materials?
3 What role does internal geometry play in determining the properties of mechanical metamaterials?
4 Explain the concept of programmable metamaterials.
5 What are adaptive metamaterials and how do they respond to external stimuli?
6 What is the importance of computational design in the development of mechanical metamaterials?
7 Explain the concept of topology optimization in metamaterial design.

8 How does generative design contribute to the development of complex metamaterial structures?
9 What role does finite element modeling play in metamaterial analysis?
10 How can machine learning accelerate metamaterial design and optimization?
11 What are smart materials and how are they integrated into metamaterial systems?
12 Explain the concept of shape-morphing metamaterials
13 What are stimuli-responsive metamaterials?
14 How do programmable metamaterials enable adaptive structural performance?
15 Discuss the importance of reconfigurable metamaterial structures.
16 What are multifunctional metamaterials?
17 How do mechanical–thermal metamaterials work?
18 Explain the concept of mechanical–acoustic metamaterials.
19 What are mechanical–electromagnetic metamaterials?
20 How do energy harvesting metamaterials contribute to sustainable technologies?
21 Why are advanced manufacturing technologies important for metamaterial fabrication?
22 What is large-scale additive manufacturing and how is it used in metamaterial production?
23 Explain the concept of multi-material fabrication in metamaterials.
24 What are hybrid fabrication techniques?
25 How do nanoscale and microfabrication techniques support metamaterial development?
26 Why is reliability an important consideration in mechanical metamaterials?
27 Explain the concept of fatigue in metamaterial structures.
28 What factors influence structural stability in metamaterials?
29 How do environmental conditions affect metamaterial performance?
30 What is the role of structural health monitoring in metamaterial systems?
31 How can metamaterials contribute to sustainable engineering practices?
32 Explain the concept of lightweight design for energy efficiency
33 What are recyclable metamaterials?
34 How do bio-inspired metamaterials improve structural efficiency?
35 What are sustainable manufacturing approaches in metamaterial production?
36 How can metamaterials improve aerospace and space structures?
37 Discuss the role of metamaterials in biomedical devices and implants.
38 How can metamaterials enhance civil infrastructure and smart city development?
39 What advantages do metamaterials provide in robotics and autonomous systems?
40 How can metamaterials contribute to energy harvesting systems?
41 What are the main manufacturing challenges associated with metamaterials?
42 Why is scalability an important issue in metamaterial production?
43 Explain the challenges related to material durability and fatigue resistance.
44 What difficulties arise in integrating multifunctional properties in metamaterials?
45 Why is standardization important for metamaterial development?

46 How can artificial intelligence support metamaterial research?
47 What are self-healing metamaterials and why are they important?
48 Explain the concept of bio-inspired metamaterial architectures.
49 What role will advance additive manufacturing play in the future of metamaterials?
50 Discuss the future outlook of mechanical metamaterials in engineering and technology.

Bibliography

Akhmetshin, L., Iokhim, K., Kazantseva, E., & Smolin, I. (2023). Influence of topological defects on the mechanical response of unit cells of the tetrachiral mechanical metamaterial. *Designs*, *7*(6), 129 https://doi.org/10.3390/designs7060129

Antoniadis, I., Chronopoulos, D., Spitas, V., & Koulocheris, D. (2015). Hyper-damping properties of a stiff and stable linear oscillator with a negative stiffness element. *Journal of Sound and Vibration*, *346*, 37–52. https://doi.org/10.1016/j.jsv.2015.02.028

Bonfanti, S., Hiemer, S., Zulkarnain, R., Guerra, R., Zaiser, M., & Zapperi, S. (2024). Computational design of mechanical metamaterials. *Nature Computational Science*, *4*(8), 574–583. https://doi.org/10.1038/s43588-024-00672-x

Buriak, I. A., Zhurba, V. O., Vorobjov, G. S., Kulizhko, V. R., Kononov, O. K., & Rybalko, O. (2016). Metamaterials: Theory, classification and application strategies. *Journal of Nano- and Electronic Physics*, *8*(4(2)), 04088-1. https://doi.org/10.21272/jnep.8(4(2)).04088

Cerniauskas, G., Sadia, H., & Alam, P. (2024). Machine intelligence in metamaterials design: a review. *Oxford Open Materials Science*, *4*(1), itae001. https://doi.org/10.1093/oxfmat/itae001

Chen, X., Laforge, N., Ji, Q., Tan, H., Liang, J., Ulliac, G., … Kadic, M. (2020). Introduction to mechanical metamaterials and their effective properties. *Comptes Rendus Physique*, *21*(7–8), 751–765. https://doi.org/10.5802/crphys.30

Cuan-Urquizo, E., & Guerra Silva, R. (2023). Fused filament fabrication of cellular, lattice and porous mechanical metamaterials: a review. *Virtual and Physical Prototyping*, *18*(1), e2224300. https://doi.org/10.1080/17452759.2023.2224300

Dudek, K. K., Kadic, M., Coulais, C., & Bertoldi, K. (2025). Shape-morphing metamaterials. *Nature Reviews Materials*, *10*(10), 783–798. https://doi.org/10.48550/arXiv.2501.14804

Fok, L., Ambati, M., & Zhang, X. (2008). Acoustic metamaterials. *MRS Bulletin*, *33*(10), 931–934. https://doi.org/10.1557/mrs2008.202

Gariya, N., Shaikh, A., Singh, H., & Chamoli, S. (2026). A comprehensive analysis of mechanical metamaterial types and applications along with challenges and current emerging trends in research. *Journal of Research and Applications in Mechanical Engineering*, *14*(1). https://ph01.tci-thaijo.org/index.php/jrame/article/view/260477

He, J., Wang, Y., Shen, Z., Xia, L., & Xiong, Y. (2024). Assembled mechanical metamaterials with integrated functionalities of programmable multistability and multitransition behaviors. *Materials Horizons*, *11*(24), 6371–6380. https://doi.org/10.1039/D4MH00906A

Humnabad, P. S., Tarun, R., & Das, I. N. D. R. A. N. E. E. L. (2022). An overview of direct metal laser sintering (DMLS) technology for metal 3D printing. *Journal of Mines, Metals and Fuels*, *70*, 127–133. https://doi.org/10.18311/jmmf/2022/30681

Imediegwu, C., Clarke, D., Carter, F., Grimm, U., Jowers, I., & Moat, R. (2023). Mechanical characterisation of novel aperiodic lattice structures. *Materials & Design*, *229*, 111922. https://doi.org/10.1016/j.matdes.2023.111922

Iyer, A. K., Alu, A., & Epstein, A. (2020). Metamaterials and metasurfaces—Historical context, recent advances, and future directions. *IEEE Transactions on Antennas and Propagation*, *68*(3), 1223–1231. https://doi.org/10.1109/TAP.2020.2969732

Jang, J. Y., Park, C. S., & Song, K. (2022). Lightweight soundproofing membrane acoustic metamaterial for broadband sound insulation. *Mechanical Systems and Signal Processing*, *178*, 109270. https://doi.org/10.1016/j.ymssp.2022.109270

Jiao, P., Mueller, J., Raney, J. R., Zheng, X., & Alavi, A. H. (2023). Mechanical metamaterials and beyond. *Nature Communications*, *14*(1), 6004. https://doi.org/10.1038/s41467-023-41679-8

Khosravani, M. R., Anders, D., Ayatollahi, M. R., & Reinicke, T. (2025). Fabrication of mechanical metamaterials by 3D printing: recent advancements and current challenges. *Archives of Civil and Mechanical Engineering*, *25*(5), 244. https://doi.org/10.1007/s43452-025-01290-8

Korner, C. (2016). Additive manufacturing of metallic components by selective electron beam melting—a review. *International Materials Reviews*, *61*(5), 361–377. https://doi.org/10.1080/09506608.2016.1176289

Li, C., Peng, Z. K., & He, Q. (2022). Stimuli-responsive metamaterials with information-driven elastodynamics programming. *Matter*, *5*(3), 988–1003. https://doi.org/10.1016/j.matt.2021.11.031

Lim, T. C. (2015). *Auxetic materials and structures* (Vol. 33). Singapore: Springer Singapore. https://doi.org/10.1007/978-981-287-275-3_1

Montemayor, L., Chernow, V., & Greer, J. R. (2015). Materials by design: Using architecture in material design to reach new property spaces. *MRS Bulletin*, *40*(12), 1122–1129. https://doi.org/10.1557/mrs.2015.263

Norris, A. N. (2006). Poisson's ratio in cubic materials. *Proceedings of the Royal Society A: Mathematical, Physical and Engineering Sciences*, *462*(2075), 3385–3405. https://doi.org/10.1098/rspa.2006.1726

Phoenix, A. A., & Wilson, E. (2018). Adaptive thermal conductivity metamaterials: Enabling active and passive thermal control. *Journal of Thermal Science and Engineering Applications*, *10*(5), 051020. https://doi.org/10.1115/1.4040280

Reda, H., Karathanasopoulos, N., Elnady, K., Ganghoffer, J. F., & Lakiss, H. (2018). Mechanics of metamaterials: an overview of recent developments. *Advances in Mechanics of Microstructured Media and Structures*, 273–296. https://doi.org/10.1007/978-3-319-73694-5_16

Sharma, D., & Hiremath, S. S. (2022). Additively manufactured mechanical metamaterials based on triply periodic minimal surfaces: Performance, challenges, and application. *Mechanics of Advanced Materials and Structures*, *29*(26), 5077–5107. https://doi.org/10.1080/15376494.2021.1948151

Siegmund, T., & Barthelat, F. (2019). Architectured Materials Mechanics. *Journal of Applied Mechanics*, *86*(11), 110301. https://doi.org/10.1115/1.4044680

Song, J., Yan, J., & Yi, B. (2024). Design, fabrication, and characterization of hierarchical mechanical metamaterials. *Frontiers of Mechanical Engineering*, *19*(1), 3. https://doi.org/10.1007/s11465-023-0776-9

Suhas, P., Quadros, J. D., Mogul, Y. I., Mohin, M., Aabid, A., Baig, M., & Ahmed, O. S. (2025). A review on mechanical metamaterials and additive manufacturing techniques for biomedical applications. *Materials Advances*, *6*(3), 887–908. https://doi.org/10.1039/D4MA00874J

Surjadi, J. U., Gao, L., Du, H., Li, X., Xiong, X., Fang, N. X., & Lu, Y. (2019). Mechanical metamaterials and their engineering applications. *Advanced Engineering Materials*, *21*(3), 1800864. https://doi.org/10.1002/adem.201800864

Walser, R. M. (2001). Electromagnetic metamaterials. *Complex Mediums II: Beyond Linear Isotropic Dielectrics*, *4467*, 1–15. https://doi.org/10.1117/12.432921

Wang, L., Chan, Y. C., Ahmed, F., Liu, Z., Zhu, P., & Chen, W. (2020). Deep generative modeling for mechanistic-based learning and design of metamaterial systems. *Computer Methods in Applied Mechanics and Engineering*, *372*, 113377. https://doi.org/10.1016/j.cma.2020.113377

Wang, P., Sun, Y., Zhang, Y., Wang, H., Zhang, Y., Xiao, C., … Zhou, H. (2025). Programmable wire metamaterials for visible and self-adaptive infrared camouflage. *Advanced Materials*, *37*(30), 2503587. https://doi.org/10.1002/adma.202503587

Wu, J., Zhang, Y., Yang, F., Jiang, F., Xu, X., Tan, Y., & Su, L. (2023). A hybrid architectural metamaterial combing plate lattice and hollow-truss lattice with advanced mechanical performances. *Additive Manufacturing*, *76*, 103764. https://doi.org/10.1016/j.addma.2023.103764

Xiao, R., Ding, M., Wang, Y., Gao, L., Fan, R., & Lu, Y. (2021). Stereolithography (SLA) 3D printing of carbon fiber-graphene oxide (CF-GO) reinforced polymer lattices. *Nanotechnology*, *32*(23), 235702. https://doi.org/10.1088/1361-6528/abe825

Yan, L., Guo, Z., Chen, Y., & Liu, Y. (2026). Functionally graded elastic valley metamaterials for low-frequency surface wave manipulation. *Mechanics of Advanced Materials and Structures*, *33*(1), 2433695. https://doi.org/10.1080/15376494.2024.2433695

Yuan, S., Chua, C. K., & Zhou, K. (2019). 3D-printed mechanical metamaterials with high energy absorption. *Advanced Materials Technologies*, *4*(3), 1800419. https://doi.org/10.1002/admt.201800419

Zhang, P., Qi, D., Xue, R., Liu, K., Wu, W., & Li, Y. (2021). Mechanical design and energy absorption performances of rational gradient lattice metamaterials. *Composite Structures*, *277*, 114606. https://doi.org/10.1016/j.compstruct.2021.114606

Zhang, Y., Jiang, W. Z., Jiang, W., Zhang, X. Y., Dong, J., Xie, Y. M., … Ren, X. (2025). Recent advances of auxetic metamaterials in smart materials and structural systems. *Advanced Functional Materials*, *35*(23), 2421746. https://doi.org/10.1002/adfm.202421746

Zhang, Y., Wang, L., Cai, Q., & Gou, X. (2024). A novel finite element design considering surface effects to evaluate the strength of nanomechanical metamaterials for naval architecture. *Ocean Engineering*, *311*, 118899. https://doi.org/10.1016/j.oceaneng.2024.118899

Zhao, A., Yu, W., Ye, J., Hu, Z., Gao, Y., Cheng, F., Gu, G., Li, S., Li, W., & Fang, H. (2024). Design and verifications of three building acoustic metamaterials for simultaneous noise insulation and ventilation. *Construction and Building Materials*, *456*, 139316. https://doi.org/10.1016/j.conbuildmat.2024.139316

Zheng, Y., Wang, Y., Lu, X., Liao, Z., & Qu, J. (2020). Evolutionary topology optimization for mechanical metamaterials with auxetic property. *International Journal of Mechanical Sciences*, *179*, 105638. https://doi.org/10.1016/j.ijmecsci.2020.105638

Index

Pages in **bold** refer to tables.

For Product Safety Concerns and Information please contact our EU representative GPSR@taylorandfrancis.com
Taylor & Francis Verlag GmbH, Kaufingerstraße 24, 80331 München, Germany

www.ingramcontent.com/pod-product-compliance
Ingram Content Group UK Ltd.
Pitfield, Milton Keynes, MK11 3LW, UK
UKHW021704300926
475953UK00017B/50

* 9 7 8 1 0 4 1 2 1 8 3 7 1 *